Communications in Computer and Information Science 688

Commenced Publication in 2007
Founding and Former Series Editors:
Alfredo Cuzzocrea, Dominik Ślęzak, and Xiaokang Yang

More information about this series at http://www.springer.com/series/7899

Quan Yu (Ed.)

Space Information Networks

First International Conference, SINC 2016
Kunming, China, August 24–25, 2016
Revised Selected Papers

 Springer

Editor
Quan Yu
Institute of China Electronic Equipment
Beijing
China

ISSN 1865-0929   ISSN 1865-0937 (electronic)
Communications in Computer and Information Science
ISBN 978-981-10-4402-1   ISBN 978-981-10-4403-8 (eBook)
DOI 10.1007/978-981-10-4403-8

Library of Congress Control Number: 2017936707

Printed on acid-free paper

This Springer imprint is published by Springer Nature
The registered company is Springer Nature Singapore Pte Ltd.
The registered company address is: 152 Beach Road, #21-01/04 Gateway East, Singapore 189721, Singapore

Preface

SINC 2016 was the First Space Information Networks Conference of the Department of Information Science, National Natural Science Foundation of China. SINC is supported by the key research project on the basic theory and key technology of space information networks of the National Natural Science Foundation of China, and organized by the "space information network" major research program guidance group. The aim is to explore new progress and developments in space information networks and related fields, to show the latest technology and academic achievements in space information networks, to build an academic exchange platform for researchers at home and abroad working on space information network and industry sectors, to share the achievements and experience of research and applications, and to discuss new theory and new technologies in space information networks. There were three tracks in SINC 2016: models of space information networks and mechanisms of high-performance networking, theory methods of high-speed transmission in space dynamic networks, and sparse representation and fusion processes in space information.

This year, SINC received 139 submissions, including 97 English papers and 42 Chinese papers. After a thorough reviewing process, 24 outstanding English papers were selected for this volume (retrieved by EI), accounting for 24.7% of the total number of English papers, with an acceptance rate of 24.5%. This volume contains the 18 English full papers presented at SINC 2016 and six short papers.

The high-quality program would not have been possible without the authors who chose SINC 2016 as a venue for their publications. We are also very grateful to the Program Committee members and Organizing Committee members, who put a tremendous amount of effort into soliciting and selecting research papers with a balance of high quality and new ideas and novel applications.

We hope that you enjoy reading the proceedings of SINC 2016.

November 2016   Quan Yu

Organization

SINC 2016 was organized by the Department of Information Science, National Natural Science Foundation of China, Kunming University of Science and Technology, Yunnan Communication Institute, and PT Press.

Organizing Committee

General Chair

Quan Yu	Institute of China Electronic Equipment System Engineering Corporation, China
Jianya Gong	Wuhan University, China
Jianhua Lu	Tsinghua University, China

Steering Committee

Hsiao-Hwa Chen	National Cheng Kung University, Taiwan, China
George K. Karagiannidis	Aristotle University of Thessaloniki, Greece
Xiaohu You	Southeast University, China
Dongjin Wang	University of Science and Technology of China, China
Jun Zhang	Beihang University, China
Haitao Wu	Chinese Academy of Sciences, China
Jianwei Liu	Beihang University, China
Zhaotian Zhang	National Nature Science Foundation of China, China
Xiaoyun Xiong	National Nature Science Foundation of China, China
Zhaohui Song	National Nature Science Foundation of China, China
Ning Ge	Tsinghua University, China
Feng Liu	Beihang University, China
Mi Wang	Wuhan University, China
ChangWen Chen	The State University of New York at Buffalo, USA
Ronghong Jin	Shanghai Jiao Tong University, China

Technical Program Committee

Jian Yan	Tsinghua University, China
Min Sheng	Xidian University, China
Junfeng Wang	Sichuan University, China
Depeng Jin	Tsinghua University, China
Hongyan Li	Xidian University, China
Qinyu Zhang	Harbin Institute of Technology, China

Qingyang Song	Northeastern University, China
Lixiang Liu	Chinese Academy of Sciences, China
Weidong Wang	Beijing University of Posts and Telecommunications, China
Chundong She	Beijing University of Posts and Telecommunications, China
Zhihua Yang	Harbin Institute of Technology, Shenzhen, China
Minjian Zhao	Zhejiang University, China
Yong Ren	Tsinghua University, China
Yingkui Gong	University of Chinese Academy of Sciences, China
Xianbin Cao	Beihang University, China
Chengsheng Pan	Dalian University, China
Shuyuan Yang	Xidian University, China
Xiaoming Tao	Tsinghua University, China

Organizing Committee

Chunhong Pan	Chinese Academy of Sciences, China
Yafeng Zhan	Tsinghua University, China
Liuguo Yin	Tsinghua University, China
Jinho Choi	Gwangju Institute of Science and Technology, South Korea
Yuguang Fang	University of Florida, USA
Lajos Hanzo	University of Southampton, UK
Jianhua He	Aston University, UK
Y. Thomas Hou	Virginia Polytechnic Institute and State University, USA
Ahmed Kamal	Iowa State University, USA
Nei Kato	Tohoku University, Japan
Geoffrey Ye Li	Georgia Institute of Technology, USA
Jiandong Li	Xidian University, China
Shaoqian Li	University of Electronic Science and Technology of China, China
Jianfeng Ma	Xidian University, China
Xiao Ma	Sun Yat-sen University, China
Shiwen Mao	Auburn University, USA
Luoming Meng	Beijing University of Posts and Telecommunications, China
Joseph Mitola	Stevens Institute of Technology, USA
Sherman Shen	University of Waterloo, Canada
Zhongxiang Shen	Nanyang Technological University, Singapore
William Shieh	University of Melbourne, Australia
Meixia Tao	Shanghai Jiao Tong University, China
Xinbing Wang	Shanghai Jiao Tong University, China
Feng Wu	University of Science and Technology of China, China
Jianping Wu	Tsinghua University, China

Contents

The Model of Space Information Network and Mechanism of High Performance Networking

An Architecture of Space Information Networks Based-on Hybrid
Satellite Constellation   3
   Yun Jia, Jingshi Shen, and Mingrui Xin

Software Defined Integrated Satellite-Terrestrial Network: A Survey   16
   *Ye Miao, Zijing Cheng, Wei Li, Haiquan Ma, Xiang Liu,
and Zhaojing Cui*

The Development of Spacecraft Electronic System   26
   Zongfeng Ma, Lichao Sun, and Zhichao Qu

High Efficient Scheduling of Heterogeneous Antennas in TDRSS Based
on Dynamic Setup Time   35
   Lei Wang, Linling Kuang, Huiming Huang, and Jian Yan

Research on Spatial Information Network System Construction
and Validation Technology.   45
   Hongbin Zhou, Jingchao Wang, and Jincan Liu

Microwave-Laser Integrated Network Component Technology   54
   Xiujun Huang, Dele Shi, and Zongfeng Ma

Confer on the Challenges and Evolution of National Defense Space
Information Network for Military Application.   66
   *Yong Jiang, Shanghong Zhao, Shihong Zhou, Yongjun Li,
Yongxing Zheng, and Haiyan Zhao*

Awareness of Space Information Network and Its Future Architecture   80
   Chengwu Chang, Hongyang Liu, Liyu Cheng, and Jing Fan

Theory and Method of High Speed Transmission in Space Dynamic Network

A Congestion-Based Random-Encountering Routing (CTR) Scheme   95
   Chao Zhao, Gang Xie, Jingchun Gao, and Kaiming Liu

Multiuser Detection for LTE Random Access in MTC-Oriented Space
Information Networks   104
   Qiwei Wang, Guangliang Ren, and Jueying Wu

Research Progresses and Trends of Onboard Switching
for Satellite Communications 117
   Wanli Chen, Kai Liu, and Xiang Chen

Contact Graph Routing with Network Coding for LEO
Satellite DTN Communications.................................. 126
   Cuiqin Dai, Qingyang Song, Lei Guo, and Qianbin Chen

A Trust Holding Based Secure Seamless Handover in Space
Information Network... 137
   Zhuo Yi, Xuehui Du, Ying Liao, and Lifeng Cao

Modulation Index Selection Strategy for Quasi-Constant Envelope OFDM
Satellite System 151
   Cheng Wang, Yizhou He, Gaofeng Cui, and Weidong Wang

Wide-Angle Scanning Phased Array Antenna for the Satellite
Communication Application 162
   Chunmei Liu and Shaoqiu Xiao

A Movable Spot-Beam Scheduling Optimization Algorithm
for Satellite Communication System with Resistance to Rain Attenuation.... 172
   *Houtian Wang, Xingpei Lu, Dong Chen, Yufei Shen, Ying Tao, Zihe Gao,
and Guoli Wen*

End-to-End Stochastic QoS Performance Under Multi-layered
Satellite Network 182
   *Min Wang, Xiaoqiang Di, Yuming Jiang, Jinqing Li, Huilin Jiang,
and Huamin Yang*

A Non-stationary 3-D Multi-cylinder Model for HAP-MIMO
Communication Systems 202
   Zhuxian Lian, Lingge Jiang, and Chen He

Sparse Representation and Fusion Process in Space Information

Double Layer LEO Satellite Based "BigMAC" Space Information
Network Architecture.. 217
   Keke Zhang, Lei Xia, Shengyu Zhang, Chaoming Si, and Shilong Zhou

The Assumption of the TT&C and Management for SIN Based on TDRS
SMA System 232
   Liang Zhu, Huiming Huang, Jian Gao, and Bin Luo

Architecture and Application of SDN/NFV-enabled Space-Terrestrial
Integrated Network 244
   Xiangyue Huang, Zhifeng Zhao, Xiangjun Meng, and Honggang Zhang

Service Customized Software-Defined Airborne Information Networks   256
  Xiang Wang, Shanghong Zhao, Jing Zhao, Yongjun Li, Haiyan Zhao,
  and Yong Jiang

Design and Application Analysis of the Global Coverage Satellite System
for Space Aeronautics ATM Information Collection   266
  Changchun Chen, Zhengquan Liu, Wei Fan, and Tao Ni

A Distributed Algorithm for Self-adaptive Routing in LEO
Satellite Network   274
  Hao Cheng, Meilin Liu, Songjie Wei, and Bilei Zhou

Author Index   287

The Model of Space Information Network and Mechanism of High Performance Networking

An Architecture of Space Information Networks Based-on Hybrid Satellite Constellation

Yun Jia[✉], Jingshi Shen, and Mingrui Xin

Research and Development Center, Shandong Institute of Aerospace
Electronics Technology, Yantai, China
Jiayun0223@126.com

Abstract. In this paper, based-on the GEO/IGSO satellite, we design a constellation to achieve global coverage. And the optimization for the coverage performance and inter-satellite link performance is provided. In order to maintain connection with ground station, two methods is proposed. The simulations and analysis for the constellation are given. The constellation can achieve the requirement of coverage and communication.

Keywords: Space information networks · Satellite networks · Hybrid satellite constellation · Network architecture · Constellation design

1 Introduction

The development of space technology, communication technology and sensor technology make the networks becoming three-dimensional and integration. Connecting all kinds of networks (satellites networks, ground networks etc.) together is a big problem. The space information networks can settle this problem. The space information networks can make different types of networks incorporating into a unified framework [1].

In the construction of space information networks, designing the satellite constellation is the first stage. Generally, the satellites used for providing fixed particular communication service selected GEO [2]. To achieve global communication, multiple satellites making up a constellation is the common method [3].

The coverage is the first problem need to be settled for designing a constellation. With a given elevation angle, the height is the element which influences the area of satellite coverage [4]. The satellite in GEO can provide a wide range of coverage. But global communication by non-geostationary satellite constellation is the future direction of development [5].

In this paper, based-on the GEO/IGSO satellite, we design a constellation with the optimal inter-satellite link (ISL), and then we revise the design to divide the constructing process into three stages.

© Springer Nature Singapore Pte Ltd. 2017
Q. Yu (Ed.): SINC 2016, CCIS 688, pp. 3–15, 2017.
DOI: 10.1007/978-981-10-4403-8_1

2   The Design Theory for High Earth Orbit Constellation

High Earth Orbit (HEO) includes Geostationary Earth Orbit (GEO) and Inclined Geosynchronous Satellite Orbit (IGSO). There are several advantages when the GEO satellite is used to provide regional communication service. One GEO satellite can cover 42.2% of the earth's surface, and GEO satellites have been used widely, such as Inmarsat, MSAT, N-STAR, etc. [6]. There are some week points when the GEO mobile satellite communication system operates alone. For instance, it's difficult for providing mobile service to the users living in high latitude area, and there are blind spots near the poles. These disadvantages cause a large fading margin to hold link availability. Thus, the GEO satellite needs a large antenna for connecting with mobile terminals.

The orbit altitude of IGSO is the same as that of GEO. And its orbit period is the same as the earth's rotation cycle. The satellites operating on IGSO overcome the problem of low elevation angle in high latitude regions. The orbit inclination of IGSO is over $0°$. Thus, the sub-satellite track is not a point but an "8"-shape pattern symmetrical about the equator. Moreover, with the orbit inclination increasing, the area of the "8"-shape pattern becomes large [7].

The coverage of IGSO satellite is lower than that of GEO satellite. But the coverage of IGSO satellite constellation could be better than that of GEO satellite constellation. This is because a higher average elevation angle, and a better multi-satellite coverage [8]. The IGSO constellation can provide a better link availability and diversity gain under the same fading margin when the satellite communication paths are independent of each other.

The orbit inclination, right ascension of ascending node (RAAN) and true anomaly are 3 elements for IGSO satellite constellation to achieve better coverage performance [9]. RAAN influences the longitude of satellite over the equator. In order to get better coverage performance, the sub-satellite tracks of the IGSO satellites should be the same, and the longitude is better at the center of the coverage area. In fact, the available orbit is limited.

2.1   The Distribution Standards for the Satellites in GEO/IGSO Constellation

In order to achieve global coverage, the IGSO constellation needs to cover the Polar Regions. Thus, the orbit eccentricity is near zero, and the shape of the orbit is set up. In order to get better coverage performance, the sub-satellite tracks of the IGSO satellites should be the same, and the IGSO satellites are distributed evenly. The orbit inclination, right ascension of ascending node (RAAN), argument of perigee and true anomaly are main parameters used to adjust the orbit for IGSO satellite constellation.

The orbit inclination: the angle between the surface of the orbit and the equatorial plane of the earth. The angle is measured from the east to the north at the intersection point between the equator and the orbit. The orbit inclination influences the coverage performance and the communication elevation angle. Thus, constellation design needs to select a suitable orbit inclination which meets the requirement of system performance.

RAAN, argument of perigee and true anomaly together determine the longitude of which the IGSO satellite crosses the equator. In order to achieve preferable coverage performance in a given area, more than one IGSO satellites are designed. These satellites are distributed on the same orbit evenly, and have the same RAAN and sub-satellite track whose longitude is located in the center of the particular region. The optimal ISL performance needs the satellites distributed on the same orbit to be as much as possible. The same sub-satellite tricks make the coverage performance of satellites optimal. It means that the system needs less satellites and economic cost. The complexity of the system has also been reduced.

Generally, two IGSO satellites repeat the same sub-satellite tricks when they have same orbit inclination and the result of RAAN plus initial phase. It implies that:

$$\begin{cases} i_1 = i_2 \\ \Omega_1 + \mu_{01} = \Omega_2 + \mu_{02} \end{cases} \tag{1}$$

Where, i is the orbit inclination, Ω is RAAN, μ is the initial phase of the satellite. Moreover, the two IGSO satellite located on the same orbit when their orbit inclination and RAAN satisfy:

$$\begin{cases} i_1 = i_2 \\ \Omega_1 = \Omega_2 \end{cases} \tag{2}$$

The communication elevation angle cannot be less than 15°. Then, as shown in Fig. 1, we can get that the half beam angle is 8.38°.

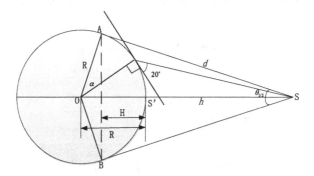

Fig. 1. Calculation method of beam angle

2.2 The Standards of ISL Establishment for GEO/IGSO Constellation

The GEO/IGSO constellation uses IGSO satellites with a large orbit inclination. In order to save the satellite performance, the satellite needs attitude control that the satellite rotates around its axis. The axis is the line which connects the center of the satellite and the core. It's difficult to build ISLs, when tracking a satellite with antenna according to azimuth and pitch. There are two methods for the attitude control problem. One method is pointing to the Earth's Core (PEC). The other method is No Attitude Control (NAC).

PEC: The antenna of satellite points to the 0 value direction which is the earth's core. Assuming that, the range of pitching and rolling is ω. Then, we can get that the antenna of satellite could coverage an area covered by a cone whose center is the satellite and angle is ω. If one satellite has a great distance with another, the ISL can build with small level of pitching and rolling.

NAC: Using a particular orbit to design the constellation. This is a unique orbit which RAAN is zero. On this orbit, there is no need for the satellites to keep attitude control.

3   Design of Hybrid Satellite Constellation

In this paper, we design a global coverage constellation scheme. The communication elevation angle of the constellation is at least 15°. The coverage performance of the constellation is more than 90% of the world, time division coverage for part of region, full coverage of low latitude region and double coverage for China and around. No more than eight satellites involve in the constellation, and each satellite connects with the earth station via no more than two satellite relay.

3.1   A Constellation Design for Optimal ISL

In order to achieve a perfectible coverage performance, the IGSO with a small orbit inclination is selected. In this scheme, 3 GEO satellites and 4 IGSO satellites are used. The parameters of these satellites are represented as follow: (Tables 1, 2, 3, 4 and 5)

Table 1. GEO parameters

GEO satellite	Geographical position
GEO1	E100°
GEO2	W130°
GEO3	E40°

Table 2. IGSO11 parameter

Satellite parameter	Value
Semimajor axis	42164.2 km
Eccentricity	0
Inclinaon	33.5°
Argument of perigee	240°
RAAN	0°
True anomaly	0°

Table 3. IGSO12 parameter

Satellite parameter	Value
Semimajor axis	42164.2 km
Eccentricity	0
Inclinaon	33.5°
Argument of perigee	0°
RAAN	0°
True anomaly	20°

Table 4. IGSO21 parameter

Satellite parameter	Value
Semimajor axis	42164.2 km
Eccentricity	0
Inclinaon	33.5°
Argument of perigee	120°
RAAN	120°
True anomaly	0°

Table 5. IGSO31 parameter

Satellite parameter	Value
Semimajor axis	42164.2 km
Eccentricity	0
Inclinaon	33.5°
Argument of perigee	0°
RAAN	24°
True anomaly	0°

This scheme satisfies requirement of the coverage for low latitude region, multiple coverage for China and around, and coverage of the Polar Regions. The global coverage performance of this constellation is shown in Fig. 2. The double coverage performance is shown in Fig. 3.

As shown in Fig. 4, more than 45% of Polar Regions are covered. Figures 5 and 6 show the coverage of low latitude region and China respectively.

In this scheme, IGSO11 and IGSO 12 is located at a same orbit. The RAAN of this orbit is zero, and the orbit inclination is 33.5°. We can get that it's easy to build an ISL. Thanks to that IGSO11 can connect with Xiamen station at any time, IGSO 12 can connect to ground station via ISL in real time. IGSO21 and IGSO31 need attitude control. ISGO 21 and IGSO31 can connect with Xiamen station in real time, because the orbit inclination is 33.5° (as shown in Fig. 7).

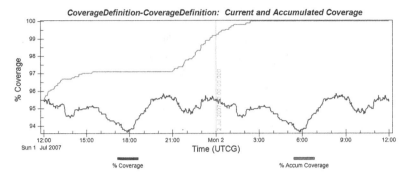

Fig. 2. The global coverage performance

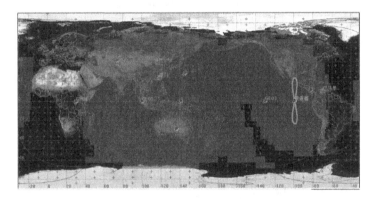

Fig. 3. The double coverage performance

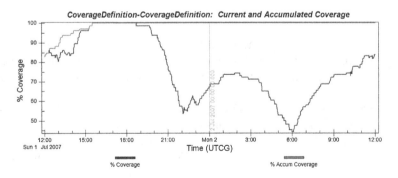

Fig. 4. The coverage of Polar Regions

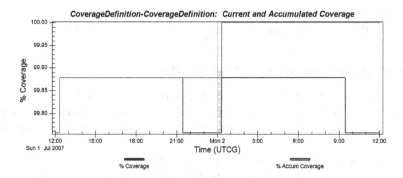

Fig. 5. The coverage of low latitude region

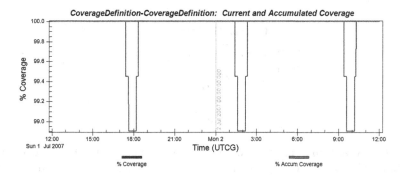

Fig. 6. Coverage of China and around

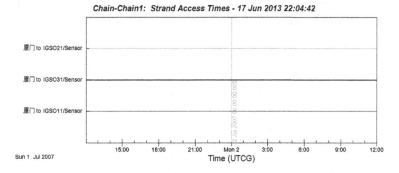

Fig. 7. The connection performance with Xiamen station

In this scheme, 3 GEO satellites and 4 IGSO satellites are used to achieve an optimal global coverage constellation design. This scheme reduces the requirement for ISL antenna with small orbit inclination of IGSO. The global coverage of world is more than 94%, the cumulative coverage is 100%. The real time coverage of the area within the range of 60° north of south latitude is close to 100%. The coverage of Polar Regions is more than 45%. The constellation can connect with Xiamen station at any time. This scheme has the minimum requirement for ISL antenna. Thus, the ISL is the best.

3.2 A Staged Implementation Constellation Design

Commonly, the process of building a constellation is a staged work, because the cost of satellite is huge. In this section, we revise the design of constellation with optimal ISL. The Staged Implementation Constellation (SIC) Design can be divided into three stages to achieve.

In the first stage of the SIC design, four GEO satellites and one IGSO satellite are selected to achieve global coverage. These four GEO satellites are used to match the requirement of covering the low latitude region and China doubled. The parameters are shown in Table 6:

Table 6. GEO parameters

GEO satellite	Geographical position
GEO1	E100°
GEO2	E160°
GEO3	W130°
GEO4	E40°

GEO1 covers China and around. GEO2 covers Western Pacific. GEO3 covers Eastern Pacific. GEO4 covers Africa. The double coverage of China and around is achieved by GEO1 and GEO4.

IGSO satellite provides time divided coverage to South America and Polar Region. RAAN of ISGO orbit is zero, and the longitude of intersection point is W80°. The parameters of IGSO are shown in Table 7.

Table 7. IGSO12 parameters

Satellite parameter	Value
Semimajor axis	42164.2 km
Eccentricity	0
Inclinaon	48.5°
Argument of perigee	20°
RAAN	0°
True anomaly	0°

In the second stage, the SIC design add an IGSO satellite (named IGSO11). IGSO11 is also located at the orbit whose RAAN is zero, and the initial phase difference between IGSO12 and IGSO11 is 140°. The parameters of IGSO11 are shown in Table 8.

Table 8. IGSO11 parameters

Satellite parameter	Value
Semimajor axis	42164.2 km
Eccentricity	0
Inclinaon	48.5°
Argument of perigee	0°
RAAN	0°
True anomaly	240°

IGSO11 and IGSO12 are in the same orbit, and they can build a stable ISL. IGSO12 can connect with ground station via this link.

In the third stage, the SIC design add two IGSO satellites with the same sub-satellite shape trick. Their parameters are shown as follow: (Tables 9 and 10)

Table 9. IGSO21 parameters

Satellite parameter	Value
Semimajor axis	42164.2 km
Eccentricity	0
Inclinaon	33.5°
Argument of perigee	120°
RAAN	120°
True anomaly	0°

Table 10. IGSO31 parameters

Satellite parameter	Value
Semimajor axis	42164.2 km
Eccentricity	0
Inclinaon	33.5°
Argument of perigee	0°
RAAN	240°
True anomaly	0°

The initial phase difference between IGSO21 and IGSO31 is 120°, and that between IGSO31 and IGSO11 is also 120°. This architecture increases the coverage of Polar Regions. Moreover, China can be covered by at least three satellites.

4   Experimental Evaluation

In this paper, based-on GEO/IGSO satellites, we design a global coverage constellation which can be achieved by multiple stages. In this section, we provide numerical simulation results to evaluate the coverage performance and the link performance of the proposed constellation schemes.

4.1   The Performance in the First Stage

Figure 8 shows the performance of low latitude region coverage. The cumulative coverage of low latitude is 100%, and real time coverage is at least 99%.

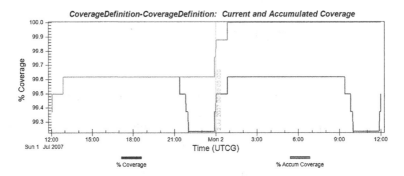

Fig. 8.   The performance of low latitude region coverage

From Fig. 9, we can get that the cumulative coverage of the world is 100% which means the constellation can provide time divided coverage to South America.

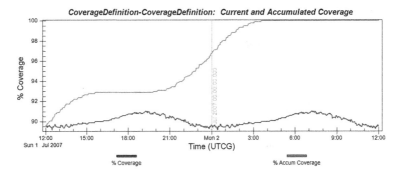

Fig. 9.   The performance of global coverage

As we mentioned above, there is no need for the satellites on GEO to keep attitude control. Thus, it's easy for the GEO satellites to build ISL. IGSO12 can use PEC method to connect with GEO2, and relay to ground station.

In the second stage, IGSO11 provides extra time divided coverage. The duration of this extra coverage is closed to 12 h. The coverage performance of the other regions is the same as that in the first stage.

Po-Facility is the area where the duration is the least. From Fig. 10, we can get that the duration of time divided coverage is more than 8 h. That means form the second stage, the duration of time divided coverage is more than 8 h for all the particular regions.

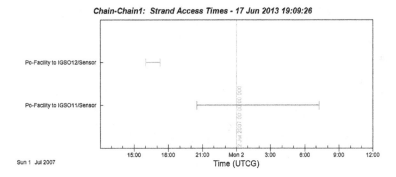

Fig. 10. The performance of time divided coverage

Adding IGSO11 to the constellation in the second stage can reduce the requirement for ISL antenna, because IGSO11 and IGSO12 can connect with each other without attitude control. But IGSO11 cannot connect with Xiamen station at any time because of the large orbit inclination, as shown in Fig. 11.

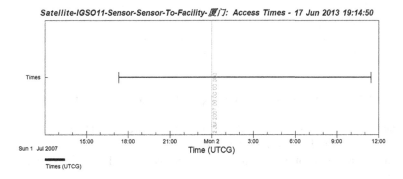

Fig. 11. The visibility of IGSO11 and Xiamen station

Otherwise, the positions of IGSO11 and GEO1 are close to each other. Thus, PEC method cannot be used to build ISL. And, tracking the Az-EI coordinates is the only way to build ISL when IGSO11 is unable to connect to the Xiamen station. During this period, pitching and phase change little.

4.2   The Performance in the Third Stage

As mentioned above, the architecture of the IGSO satellites add in the third stage can keep high elevation communication with ground station at any time. This constellation also provides triple coverage of China, double coverage of the world and real-time coverage of Polar Regions. The coverage performances of the world and the Polar Regions in the third stage are shown in Figs. 12 and 13 respectively.

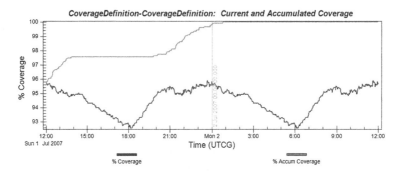

Fig. 12. The coverage performance of the world in the third stage

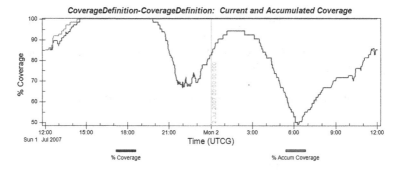

Fig. 13. The coverage performance of Polar Regions in the third stage

In the third stage, IGSO21 and IGSO31 need attitude control. Their orbit inclination is 33.5°. As shown in Fig. 14, they can connect with Xiamen station at any time with ISL. Thus, the requirement for antenna is the same as that in the second stage.

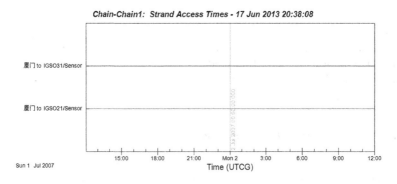

Fig. 14. The visibility of IGSO11 and Xiamen station in the third stage

The simulation analysis shows, this constellation building process can divided into multiple stages. Four GEO satellites provide full coverage of low latitude regions and then four IGSO satellites are added in different stage to cover the high latitude regions.

Together, these four GEO satellites and four IGSO satellites achieve global coverage. The cumulative coverage of the world is 100%, and the coverage of Polar Regions is more than 50%.

5   Conclusion and Future Work

In this paper, based-on the GEO/IGSO satellite, we design a constellation with the optimal ISL, and we revise the design to divide the constructing process into three stages. In order to keep connection with ground station, the PEC method and NAC method are provided. The constellation with the optimal ISL includes three GEO satellites and four IGSO satellites with small orbit inclination. This constellation can achieve the requirement of covering the entire world, Polar Regions and low latitude region. These IGSO satellites can directly connect with ground station, and the complexity for antenna is reduced. The coverage performance of the revised constellation is the same, and it supports multiple stages construction. Moreover, in the early stage of constellation construction, or when the satellite is in trouble, the revised constellation can maintain the basic communication function.

References

1. Wu, T.-Y., Wu, S.-Q.: Research on the design of orthogonal circular orbit satellite constellation. Syst. Eng. Electron. **30**(10), 1966–1972 (2008)
2. Wang, Z.: Architecture design and analysis of multi-layer satellite networks. Harbin Institute of Technology, Harbin (2007)
3. Chang-sheng, G.A.O., Wu-xing, J.: The analysis of regional navigation constellation composing eight IGSO satellites. J. Harbin Inst. Technol. **39**(7), 1036–1039 (2007)
4. Chang, H.: Study on collaborative optimization design of satellite constellation. Huazhong University of Science and Technology, Wuhan (2012)
5. Geng, L., Hu, J.H., Wu, S.Q.: Design and performance analysis of non GEO regional constellation. Sichuan Commun. Technol. **20**(02), 32–37 (2002)
6. Xiaoyu, C., Maocai, W., Guangming, D., Lei, P.: Design and implementation of satellite constellation performance estimation system. Comput. Appl. Softw. **32**(11), 44–48 (2015)
7. Wan, P., Zhang, G.: Communication protocols and operation modes of global satellite navigation constellation networks. J. Spacecr. TT&C Technol. **34**(5), 444–452 (2015)
8. Li, Y.-J., Zhao, S.-H., Wu, J.-L.: A general evaluation criterion for coverage performance of LEO constellations. J. Astronaut. **34**(4), 410–417 (2014)
9. Guo, C., Deng, L., Hu, Y.: Constellation design for regional coverage satellite. Microcomput. Appl. **30**(11), 47–49 (2014)

Software Defined Integrated Satellite-Terrestrial Network: A Survey

Ye Miao[1], Zijing Cheng[1(✉)], Wei Li[1], Haiquan Ma[1], Xiang Liu[2], and Zhaojing Cui[1]

[1] State Key Laboratory of Space-Ground Integrated Information Technology, Beijing Institute of Satellite Information Engineering, Beijing 100029, China
my_miaoye@163.com, linuxdemo@126.com, castliwei@126.com, mahaiquan@sina.com, cuizhaojing@126.com
[2] Capital Spaceflight Machinery Company, Beijing 100076, China
267180691@qq.com

Abstract. SDN paradigm has successfully manage to pave the way toward next-generation networking, but the research on SDN-based integrated satellite and terrestrial network has just started. SDN-enabled management and deployment architecture of integrated satellite-terrestrial network eases the complexity of management of infrastructures and networks, improves the maintaining and deployment costs, achieves efficient resource allocation and improves network performance of overall system. In this paper, we started introducing the SDN-based integrated satellite-terrestrial network architecture and discuss the unified and simple system functional architecture. Then we illustrate the two fundamental aspects of integrated network application functions. Following the demonstration of recent research works, we identify three challenges and discuss the emerging topics requiring further research.

Keywords: Software defined networking · Integrated satellite-terrestrial network · Network architecture

1 Introduction

The integration of satellite and terrestrial networks has been brought into research and discussed for years [1]. Back then, the satellite network has not been fully developed, the costly deployment and bandwidth resources limit it for valued uses, such as emergency response, military missions, world-wide operations and so on. With the rapid development of satellite technologies, satellite networks find the path for casual use, wide applications: communication, data transfer, remote sensing and Hi-Fi observation, and even Internet browsing. The satellite network can be integrated as one high-delay path, as complementary to terrestrial fixed and mobile access so that to increase QoS and QoE level delivered to end-users. In this context, satellite network shows that it is an essential part in the future heterogeneous networks (shown in Fig. 1): television broadcast, back-hauling of data in remote areas, instance mobile telephony, aircraft telecommunication services, and so on.

© Springer Nature Singapore Pte Ltd. 2017
Q. Yu (Ed.): SINC 2016, CCIS 688, pp. 16–25, 2017.
DOI: 10.1007/978-981-10-4403-8_2

Software defined and network virtualization technologies are also positioned as central technology enablers towards improving and more flexible integration of satellite and terrestrial systems [2]. SDN has been developed in terrestrial networks and achieved promising results. While the development of SDN-based satellite network has just started [3]. Essentially, SDN separates the data plane that only forwards packets and control plane, which is the centralized management of networks. In this case, it simplifies the connectivity of complex and heterogeneous infrastructures. It presents global network view and certain central control ability. This improves the collaboration between satellites and the compatibility of heterogeneous space systems. With all the benefits and advance SDN technology brings, how does the SDN technologies enhance the network performance and deliver high-level service quality in such integrated networks is worth investigating. Thus, in this paper, we summarize the technologies and research works have been done in each area, to discuss the research aspects, and hopefully to point out the research directions for the SDN based integrated system.

This paper investigates how SDN/NFV technologies can enhance the operation of satellite networks and the development and management of communication services across integrated satellite-terrestrial configuration variants. The advanced and newly brought-up techniques, schemes and the requirements are introduced. Besides, some challenges and possible directions are discussed in this area. The remainder of this paper is organized as follows. In Sect. 2, the network functional architecture is discussed. Section 3 illustrates the design aspects in network applications. Section 4 identifies some research challenges and points possible directions. Finally, Sect. 5 comes the conclusion.

2   Architecture of SDN-Based Integrated Satellite-Terrestrial Network

It is of utmost importance that next generation network architecture support multiple layers and heterogeneity of network technologies including satellite communications, WLANs, cellular networks and also kinds of terrestrial ad-hoc networks (shown in Fig. 1). In this integrated system, the communications happen in a wide range: communications between satellites and terrestrial, communications within terrestrial different networks, and also inter-satellite communications. SDN paradigm represents an opportunity to make it easier to deploy and manage different types of networks, including satellite networks, WLANs, and cellular networks. One of the most notably opportunities SDN technology provide is the simplification of management. The new SDN-enabled management and deployment architecture of hybrid satellite-terrestrial network eases the complexity of management of infrastructures and networks, improves the network performance of overall system, and decreases the maintaining and deployment cost. SDN-based implementation of hybrid architecture can bring the appropriate control level that current protocols and mechanisms cannot efficiently achieve.

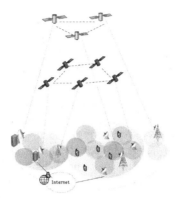

Fig. 1. Illustration of integrated satellite and terrestrial networks.

Based on the recent researched carried our, a unified functional architecture for SDN-based integrated satellite-terrestrial network is illustrated and shown in Fig. 2. Networks can be divided in three planes of functionality: the data, control, and management planes. Generally, the data plane consists of satellite and terrestrial switches and simply performing flow-based packet forwarding. Management plane includes networking applications, service interfaces, and network status management. Different network characters need to be monitored in this layer. The control plane consists of controllers located in the earth stations and terrestrial networks, which centralize all the network intelligence and perform network control for routing, handover, resource allocation and so on. Fallen within this range, research works has been focused on different aspects, and varies mainly on the design of controllers and switches.

Paper [4] introduces an SDN-enabled satellite/ADSL hybrid architecture. The SDN controller can be hosted at the service operator. In this case, the network application is running on top of the controller. Based on the data flow identifications and the designed dynamic forwarding rules, the data flow can be dispatched to the most appropriate link temporarily and dynamically to achieve its QoS requirements, with satisfying efficient utilization of different transmission links. In this case, the service provider allocates all its possessed resources including both satellite and terrestrial available bandwidth for different application requirements. However, instead of obtaining global vision, this kind of controller design achieves only partially/locally revision. Authors in [5] proposed a software-defined satellite network architecture-OpenSAN, which contains data plane, control plane and management plane. Data plane consists of multi-layered satellite infrastructures (e.g. GEO, MEO and LEO) and terminal routers distributed around the world. Control plane consists of the three GEO satellites which covers the whole data plane. The control plane GEOs communicate with terrestrial network by a centralized control center (NOCC) via a primary GEO, or by distributed NOCCs to increase the reliability. The NOCC is the management plane of the multi-layered satellite network. In this kind of architecture, the control plane-GEO group monitors the networks status (link status, network traffic, different flow status) information, and transmits the information to management plane-NOCC. Based on the various applications and schemes, NOCC runs different modules, such as routing policy calculation, virtualization, security, resource utilization and mobility management. After this, NOCC transmits the calculation results (e.g. new flow table) down to data plane. Data plane is responsible for translating the rules from management plane to data plane, and finally data plane (e.g. satellites and routes) run flow table match-action protocol and only focus on packets forwarding.

Authors in [6] proposed a new hybrid control structure with information forwarding through single layer inter-satellite links and GEO satellites broadcasting. Authors in [7] propose the integrated terrestrial-satellite software defined networking functional architecture. The paradigm shift towards virtualization of infrastructure components, pushes towards a cloud-based model for network resources and functionalities management. Control intelligence is centralized in the control layer which translates the upper layer instructions to configurations and date structures for infrastructure layer. Terrestrial and satellite network resources are federated in this layer, and the virtualized network slices are provided to the application layer users.

3   Network Functions

This section discusses the networks functions which implement control logic and dictate the behaviour of the forwarding devices. Despite the wide variety of use cases, the most essential and vital two SDN network functions are: resource management and routing mechanism.

3.1   Resource Management

The traditional resource-oriented resource management methods are no longer competitive for the largely increasing service requirements in the integrated networks. SDN-based flexible satellite resource management has been developed to advance the typical satellite broadband access service with the customer to be able to dynamically request and acquire bandwidth and QoS in a flexible and elastic manner [7]. This is to introduce more dynamicity in radio resource management of the satellite links. It optimizes the utilization of network resources, but also makes it possible to perform the network configuration, dimensioning and adjustment in real-time to fulfill the customer's expectation. Furthermore, the resource of satellite and terrestrial access networks can be federated, which means the pooling of different resources from two or more heterogeneous domains in a way to create one logical federation of network resources enabling easier control and allocation of these resources. That is to say, the network resources can be seen to transient among the networks, such as Wide Area Network (WAN), 5G networks, and satellite network for connectivity during a specific time period for service provision. This procedure can be generalized to provide the different QoS and service classes dynamically and on the fly [8]. Even in such resource federation case, the handover among different network domains and different network accesses is still necessary underlying. Authors in [9] propose a seamless handover in software defined satellite networking, but only in satellite networks.

Such SDN-based flexible federation of satellite and terrestrial networks requires efficient traffic control and traffic engineering. The main goal the traffic engineering is of minimizing power consumption, maximizing average network utilization, providing optimized load balancing and other generic traffic optimization techniques. Load balancing is one of the first to be envisioned. It distributes the traffic among the available paths/network links and among the available servers, taking into consideration of network load, link conditions, and server capacities. In this case, load balancing service alleviates the network congestions, avoids bottleneck situations, and simplifies the placement of network services in the network to provide more flexibility to network overall utilization and network servers. Thus, traffic optimization is especially important for large- scale service provider in a large integrated system, where dynamic scale-out is required. Recent work has shown that the optimizing rules placement can increase network efficiency. Traffic engineering is a crucial issue in all kinds of networks, thus, upcoming methods, techniques, and innovations can be expected in the context of SDN-based integrated satellite-terrestrial networks [10].

3.2   Routing and Networking

Routing is always the basic and important function in any network, where the priority job is to guarantee the end-to-end delivery of data packets. To achieve this goal, routing schemes are to define the path through which packets will flow from one point to another, based on network feature input. And efficient and intelligent routing protocol should be able to provide flexible adjustments for various network conditions. Diversity of network participators, the complexity and dynamic network topology raises the challenge for adaptive routing mechanism in SDN based integrated satellite-terrestrial network to achieve the internetworking within the same domain and across different network domains. In traditional integrated satellite-terrestrial networks, the interoperability of different protocols is one of the main issues, while with the SDN paradigm, the rules and regulations are the same for the overall system, components of system follow the same instruction, which erase this problem already.

The most effective aspects of routing in such system include deal with frequently changing of network topologies, and guarantee of QoS requirements of various services [11]. Firstly, the largely and highly dynamic topology changing leads to dynamic network nodes and control nodes (e.g. relatively high velocity of satellites and terrestrial terminals), which brings in much more difficulties

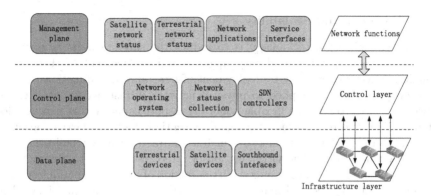

Fig. 2. Software defined integrated satellite-terrestrial networking functional architecture.

for the routing mechanisms. As the topology of both satellite and terrestrial network changes, it is difficult to maintain the stability. The static routing is clearly not suitable for such large delay network, and the dynamic routing, on the other hand, is quite resource consuming. Beside, the accusation of network status is important in such networks. The control messages, which demonstrate the network conditions, need to be delivered across different planes in SDN-based system architecture, which increases the control overhead. Hence, a scheme with trade-off between flexibility and control cost is essential for the SDN-based integrated network [6]. Secondly, one of the cases can be envisaged when taking into account Quality of Service (QoS) is that QoS oriented routing protocol. The routing mechanisms are developed to choose the best route depends on the QoS parameters and link quality. Different metrics (e.g. delay, loss rate, jitter and throughput) can be used in path selection algorithms to achieve high level of service satisfaction under specific objectives. For example, different service applications (e.g. voice call, data transfer, and video streaming) require various aspects and levels of service quality (e.g. short delay, high bandwidth and high secure). While in integrated satellite-terrestrial network, (such as GEO satellites provide long delay and world-wide transmission, LEO satellites can deliver low delay to internet browsing but costly, and the terrestrial links guarantee the low delay and probably high bandwidth). How to develop such comprehensive application-aware routing mechanisms to achieve the best use of the integrated network is of vital importance [9].

4   Ongoing Research Efforts and Challenges

This section highlights research efforts we consider of particular importance for unleashing the full potential of SDN, mainly in three aspects: flexibility and scalability, security, and performance evaluation.

4.1   Flexibility and Scalability

Network virtualization technology is to reduce the satellite network operator costs, this leads to a fast and easy upgrade and replacement of these functionalities but also flexibility to deployment of new innovative functions. Virtualization principles are applied to physical network infrastructure, abstracting network services to create a flexible pool of transport capacity that can be allocated, utilized and repurposed on demand. Essentially, network virtualization in integrated satellite-terrestrial network includes infrastructure virtualization and resource virtualization. The virtualization of radio resources is to abstract and share a number of network resources. Virtualization of network functions enables the centralized upgrade and maintenance of SDN-based architecture instead of operated on infrastructures [12]. For example, with the network virtualization paradigm, PEP (performance-enhancing proxy) will no longer be implemented as a dedicated middelbox but rather in software that can be run on different devices. In this way, the PEP function can be dedicated to a communication context (e.g. dedicated to an ST (satellite terminal)) and can be tuned according to the application requirements (security, mobility, performance, etc.). In this way, if an ST makes a handover from one satellite hub to anther, its dedicated virtual PEP will migrate to the new hub and will continue to performance the appropriate TCP optimization [3]. However, the virtualization of network functions should be developed in a unified and consistent way.

The modularity and flexibility of composition of controllers are still ongoing research. With the virtualization technique, SDN has the potential to facilitate the deployment and management of network applications and services with greater efficiency. However, SDN techniques to-date, largely target infrastructure-based networks. They promote a centralized control mechanism that is ill suited to the level of decentralization, disruption and delay present in infrastructure-less environment. In integrated satellite-terrestrial networks, the traditional centralized control mechanisms cannot be adapted and suitable for the large-scale and largely-increasing service requirements. As the essential design of SDN technology, the architecture of control plane are critical for the flexibility and scalability of integrated system.

4.2   Security

Security is the essential problem in all kinds of networks. There is a crucial need to assure the privacy and security of residents in such heterogeneous networks. Being highly programmable makes the potential impact of threads far more serious in SDN, compared to traditional networks. The research in SDN-based security is still on the early stage. Thus, security is one the top priorities in such network and more effort should be put in future researches. Possible challenges and future directions for security in SDN-based integrated satellite-terrestrial network could be classified in several groups. Firstly, some thread vectors should be identified and followed: faked or forges traffic flows in data plane, which can be used to attack forwarding devices and controllers; faulty or malicious controllers or applications in controller plane, which can be used to reprogram the entire network and grant an attacker the control of the network; lack of trusted resources for forensics and remediation, which can compromise investigations and preclude network recovery to safe condition. Secondly, orchestrating security policies across heterogeneous networks is crucial. Mechanisms, which translate security privileges across domain boundaries, are needed to enforce a uniform federated security policy in a seamless and efficient manner. Last but not least, customizing overlay networks could be used to provide secure environments [13].

4.3 Performance Evaluation

With the benefits SDN paradigm represents, a growing number of researches and experiments about SDN-based integrated satellite-terrestrial networks are expected in the near future. This will naturally create new challenges, as questions regarding SDN performance have not yet been properly investigated. Some OpenFlow based implementations have been developed for simulation studies and experimentations for the SDN-based network architecture. Except for the widely used time-consuming simulation and expensive experimental techniques for performance evaluation, analytical modeling could, in another way, draw the description of a networking architecture which paves the way for network designer to have a quick and approximate estimate of the performance of their design. Despite of the evaluation of network architecture, there are also other designed mechanisms to be evaluated. When it comes to routing mechanisms, resource allocation algorithms, and networking schemes, analytic models can quickly provide performance indicators. They can be used to capture the closed form of certain network performance, such as packet delivery rate, packet delay, buffer length, network throughput, network blocking probability and so on. Although a wide range of research works proposes SDN-based networks, there are very few performance evaluation and analytical modeling studies about these works, even for terrestrial networks, let alone for the integrated satellite - terrestrial network.

4.4 Migration and Integrated Deployment

With the benefits SDN paradigm represents, the large amount of research results, and the achievements of software-defined radio technologies, the SDN-based integrated network is reaching the migration challenge regarding with the incremental deployability. Some efforts have already devoted to the migration and hybrid SDN engineered with the current network infrastructures. The fundamental step will be allowing the coexistence of traditional environments of routers and switches with the new OpenFlow-enabled devices. Next step is to ensure the interconnection of control plane and data plane of legacy and new network elements. The initial SDN operational deployments are mainly based on virtual switch overlay models or OpenFlow based network controls. The controllers are designed to introduce SDN-like programming capabilities in traditional network infrastructures, making the integration of legacy and SDN-enabled networks a reality without side effects in terms of programmability and global network control. Future works are required to devise techniques and interaction mechanisms that maximize its inherited benefits while limiting the added complexity of the paradigm coexistence.

5   Conclusion

Traditional networks are complex and hard to manage since the control and data planes are vertically integrated. SDN creates an opportunity for solving this problem - decoupling of the control and data plane. The global view of network is logically centralized in control plane and packets delivery is highly efficient in data plane. SDN brings flexibility, automation and customization to the network, SDN paradigm represents an opportunity to make it easier to deploy and manage different types of networks, including satellite networks, WLANs, and cellular networks. SDN has successfully manage to pave the way toward next-generation networking, but the research on SDN based integrated satellite and terrestrial network has just started. In this paper, we started introducing the SDN-based integrated satellite-terrestrial network architecture and discuss the unified and simple system functional architecture. We illustrate the two fundamental aspects of integrated network applications. Following the demonstration of recent research works, we identify four challenges and discuss the emerging topics requiring further research.

Acknowledgments. This work is supported by the National Natural Science Foundation of China (No.91538202, No.91438117).

References

1. Evans, B., Werner, M., Lutz, E., Bousquet, M., et al.: Integration of satellite and terrestrial systems in future multimedia communications. IEEE Wirel. Commun. **12**, 72–80 (2015)
2. Ali, S., Sivaraman, V., Radford, A., Jha, S.: A survey of securing networks using software defined networking. IEEE Trans. Reliab. **64**(3), 1–12 (2015)
3. Yang, X., Xu, J., Lou, C.: Software-defined satellite: a new concept for space information system. In: IMCCC. IEEE (2012)
4. Bertaux, L., Medjiah, S., Berthou, P., Abdellarif, S., et al.: Software defined networking and virtualization for broadband satellite networks. IEEE Commun. Mag. **53**, 54–60 (2015)
5. Bao, J., Zhao, B., Yu, W., Feng, Z., Wu, C., Gong, Z.: OpenSAN: a software-defined satellite network architecture. In: SIGCOMM, Chicago, USA (2014)
6. Tang, Z., Zhao, B., Yu, W., Feng, Z., Wu, C.: Software defined satellite networks: Benefits and challenges. IEEE (2014)
7. Rossi, T., Sanctis, M., Cianca, E., Fragale, C., Fenech, H.: Future space-based communications infrastructures based on high throughput satellites and software defined networking. IEEE (2015)
8. Maheshwarappa, M., Bowyer, M., Bridges, C.: Software defined radio (SDR) architecture to support multi-satellite communciations. IEEE (2015)

9. Yang, B., Wu, Y., Chu, X., Song, G.: Seamless handover in software-defined satellite networking. IEEE Commun. Lett. **20**, 1768–1771 (2016)
10. Ferrus, R., Koumaras, H., Sallent, O., et al.: SDN/NFV-enabled satellite communications networks: opportunities, scenarios and challenges. Phys. Commun. **18**, 95–112 (2015)
11. Zhang, J., Gu, R., Li, H., et al.: Demonstration of BGP interworking in hybrid SPTN/IP networks. In: Asia Communications and Photonics Conference (2015)
12. Riffel, F., Gould, R.: Satellite ground station virtualization - secure sharing of ground stations using software defined networking. IEEE (2016)
13. Kreutz, D., Ramos, F.M.V.: Software-defined networking: a comprehensive survey. Proc. IEEE **103**(1), 10–13 (2014)

The Development of Spacecraft Electronic System

Zongfeng Ma[✉], Lichao Sun, and Zhichao Qu

Shandong Institute of Aerospace Electronics Technology, Yantai, China
mzf706@163.com, sunlch@163.com, quzhch@163.com

Abstract. The development of spacecraft electrical system in China has been reviewed. The electronic system has been divided into three generations of stand-alone system, federated electronic system and integrated electronic system. Then, the electrical architectures and features of the federated electronic system and the integrated electronic system have been analyzed. Finally, the development trend of spacecraft electronic system has been discussed.

Keywords: Spacecraft · Electronic system · Federated · Integrated

1 Introduction

Over the last several decades the miniaturization and integration of electronics has revolutionized many systems in various areas as diverse as computing, communications, and household appliances. Advanced systems which use less power but are able to perform more functions with a higher speed, are ideally suited for space applications. Here, we gather all the improvement and important characteristics of spacecraft electronic system. The discussed spacecrafts in this paper don't include new conceptual aircrafts, such as Fractionated Modularized Cluster Spacecraft and Micro-Satellite. Generally, spacecraft electronic system is refer to as the concept of parallel with the mechanical system. [1] And the spacecraft electronic system includes all electronic instrumentations (including the software) which carry out the non-mechanical functions, such as command, control, communication, and monitoring capabilities. The special electronic devices used for payload, or the power subsystem don't take into account in this paper.

The electronic system play an important roles onboard the spacecraft, which is the basis of the mission, function and performance of the spacecraft. The development of new electronic technologies is driven by the requirements of all areas of space applications. It can be said that the capabilities of design, testing, flight support, on orbit maintenance and space applications, are influenced greatly by the spacecraft electronic system. This paper presents the development of spacecraft electronic system from early history to new trends in China.

© Springer Nature Singapore Pte Ltd. 2017
Q. Yu (Ed.): SINC 2016, CCIS 688, pp. 26–34, 2017.
DOI: 10.1007/978-981-10-4403-8_3

2 Electronic System Evolution

Chinese strategy for supporting aerospace developments comprises a number of inter-related efforts. China aerospace has gradually developed from scratch, from small to big, from imitation to independently design, and is in a rapid development period from a big space country to a super power space country currently. The development of China spacecraft electronic system has experienced three stages over the last sixty years:

2.1 Generation 1 (Stand-Alone System)

In order to achieve the electronic functions, such as navigation and communication, the early spacecraft subsystems should be equipped with appropriative and independent sensors, processors, communication modules and other equipments for each function. All of the equipments were independent and interconnected point-to-point in the structure. All of the equipment and data are unique to a particular function and not shared with other functions. The electronic system architecture adopted by the early spacecraft which has no whole system control by the central computer is known as stand-alone system.

The instrument compartment on the first Chinese satellite Dong-Fang-Hong-1 [2] equipped with power, radar transponder for measuring track, radar beacon, telemetry device, electronic sound generator and transmitter and scientific experiment instrument had been launched in April 24, 1970. Its electronic system was stand-alone system. The specificity of the structure of the electronic system was strong. The mechanical and electronic Interface and outside dimensions were not compatible. The chassis just provided simple installation and protection for the internal electronic functional unit. Since each part of this architecture was independent, the error fault would not be spread to the other system, therefore the system had a natural feature of error isolation. However, the strong specificity and poor flexibility leaded to the difficulties in information exchange between the devices. The electromagnetic interference caused by the huge cables between devices was more serious. The reliability of the system was reduced.

2.2 Generation 2 (Federated Electronic System)

The former Ministry of Aerospace Industry Development identified the mission of China-Brazil Earth Resources Satellite (CBERS [3]) in 1986. China and Brazil jointly developed the CBERS based on the satellite of China Resources-1 in 1988. The CBERS was launched in October 14, 1999, and worked on orbit three years and ten month. The onboard data management system [3] which drew lessons from European Onboard Data Handing System (OBDH) architecture called federated electronic system, was first introduced on CBERS. The most of the satellites and spaceships followed by more than twenty years used that architecture.

The functional modules in the combined electronic system were interconnected via the digital bus (1553B). The combined electronic system achieved the functions of low bandwidth data transmission and exchanging by several processors and achieved limited share after the data transfer on the time of terminal control. Each subsystem in the combined electronic system was required to have the own controllers, sensors and actuators in order to achieve the various functions, which easily caused the entire system was extremely complex and also caused lots of unnecessary duplicate device in the system.

The combined electronic system had the main features of using time-division multiplexed data bus, standard on-board computer, standard development language and standard plug-in unit RTU. The combined electronic system was a transition of the development of spacecraft electronics. The architecture of combined electronic system was a transformational progress compared to the discrete electronic system and was the pioneer on the development to integration of the spacecraft electronics.

2.3 Generation 3 (Integrated Electronic System)

In order to improve the design and research capabilities of spacecraft, it was necessary to adopt the philosophy of integrated design and used standard interface and protocol specification between the onboard electronic equipment and therefore created an electronic system [4] which had the features of internal information share and utilization, functional integration, reorganization and optimization of resources, called *integrated electronic system*. The system achieved the functions of telemetry, remote control, energy management, thermal control, attitudes and orbit control, unlock and drive control, communications within the satellite, time management, data management, payload management, and etc. It achieved the purpose of hardware module generality, information flow rationality, functional density improvement and overall performance optimization.

3   Federated Electronic System Architecture

Because of the rapid development of the gradually improvement of electronic technology and computer technology, the onboard microprocessor had caused a fundamental change of the satellite monitoring technology since the late 1980 s. The OBDH system based on microprocessors synthesized the video portion of telemetry and remote control, but also integrated other information processing of the satellite.

Our first use of the OBDH system was on the China-Brazil Earth Resources Satellite I. The satellite platform integrated the original independent telemetry system and remote control system together to form the data management system which could achieve the functions of telemetry and remote control, but also had a certain amount of functions of self-management, control and etc.

The OBDH system had been widely used in the launched satellites and it had formed into a unified standard of China on board data management system. The data management system of spacecraft was mainly composed of the devices of central terminal unit (CTU), remote control unit, telemetry unit and several remote terminal units (RTU) and the software. The devices which were interconnected via the serial data bus completed the functions of spacecraft data transfer, processing, storage and management together.

The data management subsystem of the huge satellites and large spacecraft such as manned spaceship had the similar architecture, and the only difference was the numbers of RTUs. The difference of the data management subsystem of huge satellites and the satellite service subsystem of small spacecraft such as micro-satellite was that huge satellite used 1553B bus to connected the devices and micro-satellite used the CAN bus. The CTU of huge satellite connected plurality of RTUs, while the satellite service computer of micro-satellite connected with plurality of slave computers. Same micro-satellites embedded the RTU into the payload or control platform subsystem. The typical architecture of the data management subsystem [5] of spacecraft was shown in Fig. 1.

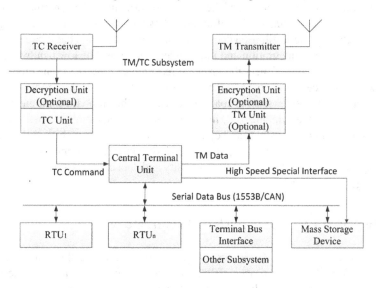

Fig. 1. The typical architecture of the data management subsystem

The main functions of the data management system of spacecraft were as follows:

① Received the upstream remote control commands and data and completed the demodulation, decoding and command verification.
② Acquired and processed the telemetry data of satellite platform and payload subsystems.
③ Provided the on-board time reference signal and managed the time.
④ Supported the information networking and interaction between the computers in the satellite.
⑤ Processed the related information of the satellite.
⑥ Provided dedicated data processing and data format for the related subsystems of the satellite.
⑦ Managed the stage of satellite and could operate in autonomous mode and external intervention mode.
⑧ Provided the mass storage function of satellite service data.

The combined electronic system was divided into a number of subsystems according to the function. Each subsystem (such as the subsystem of thermal control, attitudes and orbit control, and power) completed the dedicated control tasks by their own computer, and a CTU which was specially configured responded for management of the whole system. That system not only held relatively independent subsystems, but also had the feature of centralized and unified management of whole system.

The combined electronic system achieved the purpose of reduce the size and weight of whole system through the integrated use of a variety of techniques, provided a great solution for the processing and control of the system, greatly promoted the integration of the electronic system and enhanced the system performance. The combined electronic system achieved data sharing on the information level, but didn't achieve the integration of signal. Each subsystem still used dedicated software system and hardware equipment and had low degree of integration and independent function. The low bandwidth of the data bus gradually couldn't meet the growing requirements for data transmission. The system which was centrally controlled by a bus controller was lack of robustness.

The data management system of spacecraft based on the on-board computer had been moving towards to the miniaturized, integrated and modular direction with the rapid development of microelectronics and computer technology. Meanwhile, the increasing demand for space missions also putted forward higher requirements for the data processing capability, scalability and adaptability and data transmission standards of the data management system.

4   Integrated Electronic System Architecture

The "integrated electronics" (Avionics) was a generic term of the electronic systems of the spacecraft, and was the spacecraft's brain and nerves. The avionics system is an integrated electronic system which was based on the management of central computer, adopted the hierarchical distributed network architecture as the system architecture, completed in orbit scheduling and integrated information processing, managed and controlled of each task which was running on the satellite efficiently and reliably, monitored the state of entire satellite, coordinated the works of entire satellite, managed the payload, and achieved the unified processing and sharing of the information within the entire satellite and even the entire constellation. The system could take into account of the different requirements of the existing standard and future development electronic devices by using standard interfaces between systems.

The design of the integrated electronic system of spacecraft broke the boundaries of traditional devices and subsystems. The architecture was design according to platform information flow and energy flow and was divided into the standard modules which would be combined to the device reference to mission requirements. The system achieved the integration of information, software and hardware, and significantly improved functional density.

The traditional spacecraft platform was divided into different subsystems according to the function. The subsystems mainly included monitoring, control and communication subsystem, thermal control subsystem, power subsystem, institutional structure subsystem, attitudes and orbit control and propulsion subsystem and other subsystems. The function of each subsystem was implemented by specific device. Compared to previous system, the biggest difference was that the integrated electronic system of spacecraft emphasized that all components were placed into a complete and reasonable architecture and adopted the systems engineering approach of top-down to complete system development.

The integrated electronic system of spacecraft established a unified common architecture, divided the system into the standard modules which would be combined to the device reference to mission requirements and significantly improved the system performance and functional density (Fig. 2).

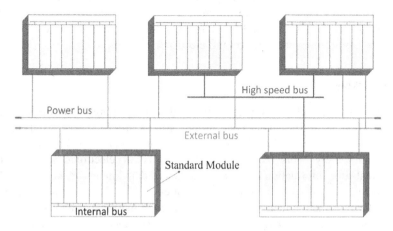

Fig. 2. The architecture of avionics

As shown in the above figure, the avionics system used the distributed and modular architecture and usually composed by a satellite management unit and several integrated business units which were connected through the external bus. The architecture of that electronic system had great flexibility and scalability of configuration and could flexibly equip with the functional modules according to different mission requirements. Making the mature functional modules into standardized shelf-type products used by different projects could shorten the development cycle, reduce the development cost and improve the system reliability.

5   The Development Trend of Spacecraft Electronic System

The current integrated electronic system has significantly improved in the respects of the architecture, the capacity of data processing, transport and storage, functional integration, standardization and modularity and etc. However, the system is insufficient in the respects of the autonomous intelligent, the dynamic reconfiguration and configuration in orbit of the system function, the comprehensive of RF and sensor, and etc. The development trend of the future spacecraft electronic system has the following aspects.

5.1 Open Architecture

The open integrated electronic system can achieve the portability of software applications and operators, complete the application interoperability between the multi-nodes within the spacecraft platform network, implement the global security of the system through adopting appropriate standard architecture and appropriate software and hardware in the embedded computer environment. Using Hierarchical division method and standard interface between the layers make it easier to the interconnection, intercommunication and interoperability of system components, migration and reuse of the hardware and software, enhancement and expansion of the system function. The standard hardware modules and the standard software functions that are summarized and optimized from the same functions of the different areas and mission tasks form the core of the system and the uniform standards, specifications and design resources database, and turn into the base framework of spacecraft electronic systems.

5.2 Networking

The future aerospace applications present the trend of architecture and network. A single spacecraft works as an intelligent node in the network and each node exchanges the information through the network. Multiple nodes fly in formation [6] and complete the mission coordinately. This requires that the spacecraft electronic systems could exchange real-time information through the advanced data link networking technology and complete the tasks of the situational awareness, the information sharing and exchange, the information fusion [7], the distribution of intelligence and command and tec. This trend will lead the spacecraft electronics technology to develop in the direction of network which is targeted the information superiority, build an integration network which is global coverage of sea and air and space, and provide the rasterized information service.

5.3 Autonomous Intelligent

The intelligent planning and scheduling is one of the key technologies to achieve the autonomous spacecraft management. The electronic system plays a crucial role as a decision-making center. The electronic system will efficiently schedule all of the in-orbit hardware and software resources according to the specified flight mission. And it will reduce the dependence of the flight control center on the ground and improve the ability to complete the mission.

5.4 Software Defined

The development trends of spacecraft are the standard Interface, the modular hardware and the dynamically configured function. The spacecraft hardware architecture will be highly uniform. The specific application of the electronic systems will be defined by the software running on them. That will achieve the system reconfiguration and function redefinition of the spacecraft on board. The "Software Configured Electronic System" is the basis that allows the spacecraft to achieve that transformation. The electronic system constitutes a software communication operating environment which may cause the developments of the software applications and the underlying hardware and software separate completely through the core framework and middleware. The application software completes the required function by using a variety of standard interfaces provided by the operating environment to invoke the underlying hardware and software resources.

5.5 Integration of the Platform and Payload

The platform and payload of the current spacecraft are mostly developed separately and they are relatively independent. There is only a small amount of data exchange between the platform and payload. With the development of technology, the integration and optimized design of the platform and payload have become an important development direction. The functions of the platform and payload can be performed by the electronic system based on the same hardware. The electronic system can provide the necessary capacity of computing, storage and processing for the functions to be achieved by loading different application software. So the designers of the payload can mainly focus on the payload performance improvement and the payload data processing algorithms implement and the designers of electronic system can focus on the development of highly configurable, high performance and general purpose platform.

6   Conclusion

This paper presented an overview of the architecture development rather than a detailed implementation of spacecraft electrical system. With the space application requirements of the spacecraft continually increased, the electronic system plays a more important role in the spacecraft. The current spacecraft electronic system has already could not meet the user expectation and the expectation will birth a profound change in aerospace electronics technology. For future work, we intend to design a soft defined next generation space avionics system that provides open architecture, flexibility within implementation, networking, third party participation, and a layered approach to both hardware and software modules that utilizes Time, Space, and I/O partitioning to realize a highly reliable and available fault tolerant system.

Acknowledgment.   In this paper, the research work was supported by National Natural Science Foundation of China under Grant (Project No. 61302162).

References

1. Li, Y.: A few key trends in foreign spacecraft electronic systems. Spacecraft Eng. **23**(6), 1–6 (2014)
2. Chen, S.: The breakthroughs in key technologies of structure design, assembly and testing of DFH-1 satellite. Spacecraft Eng. **32**(2), 127–129 (2015)

3. Zhang, Q.J., Ma, S.: The achievements and development of CBERS. Spacecraft Eng. **18**(4), 1–8 (2009)
4. Zhu, W.: System Design of Spacecraft Avionic. Shanghai Jiao Tong University, pp. 1–2 (2013)
5. Li, W., Liu, Q., Li, Y., Sun, J.: The Research of Spacecraft Data Management System. Science and Technology Committee of CAST Computers and Control Professional Group papers set (2011)
6. Huang, Y., Li, X., Wang, Z., Li, Z.: Relative position adaptive cooperative control for satellites formation flying. J. Astronaut. **35**(12), 1412–1421 (2014)
7. Zhu, J., You, M.: Reconnaissance effectiveness analyses of ocean surveillance satellite based on fusing multidimensional information. J. Syst. Simul. **26**(11), 2682–2691 (2014)

High Efficient Scheduling of Heterogeneous Antennas in TDRSS Based on Dynamic Setup Time

Lei Wang[1,3(✉)], Linling Kuang[2], Huiming Huang[3], and Jian Yan[2]

[1] School of Aerospace Engineering, Tsinghua University, Beijing 100084, China
wangleibsir@163.com
[2] Tsinghua Space Center, Tsinghua University, Beijing 100084, China
[3] Beijing Space Information Relay and Transmission Technology Center, Beijing 100094, China

Abstract. Setup time between missions in the tracking and data relay satellite system (TDRSS) was treated to be time-invariant in most previous work on the system's scheduling problem. This paper considers this problem based on dynamic setup time that is both sequence-dependent and time-dependent. Firstly, considering the actual daily operation property of the TDRSS, the original scheduling problem is transformed into the pointing route problem of inter-satellite link antennas. Then, by defining dynamic setup time and introducing new time-space constraints into the conventional scheduling model, we construct one mixed integer programming model with heterogeneous antennas. To solve this model, one algorithm based on reactive greedy randomized adaptive search procedures (Reactive GRASP) is implemented. The validity of the model and the algorithm procedure is verified through numerical simulations under different problem scales and multi-type data sets. Results show that compared with the conventional method, the scheduling success rate increases associated with an evident decline in the invalid resource consumption rate of Single Address antennas, which effectively improve the TDRSS efficiency.

Keywords: Space information network · Tracking and data relay satellite system (TDRSS) · Inter-satellite link antenna (ILA) · Scheduling · Dynamic setup time

1 Introduction

The tracking and data relay satellite system (TDRSS) is usually deployed in the geostationary orbit, which is an important component of space information network for the high-speed transmission of space information. Traditional earth-based station has less than 3% orbit coverage for a 500 km circular orbiting satellite, while a space-based relay satellite has an orbit coverage of more than 40%. And three relay satellites that are uniformly distributed in the geostationary orbit can provide 100% orbit coverage for low-orbital spacecrafts.

© Springer Nature Singapore Pte Ltd. 2017
Q. Yu (Ed.): SINC 2016, CCIS 688, pp. 35–44, 2017.
DOI: 10.1007/978-981-10-4403-8_4

In addition, the TDRSS can enhance service timeliness significantly [1–4]. The typical operational scenario of the TDRSS with two relay satellites provided by NASA is as shown in Fig. 1.

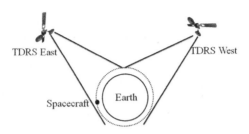

Fig. 1. The typical operational scenario of the TDRSS with two relay satellites

Relay services are processed by two types of inter-satellite link antennas (ILAs) equipped on each TDRS, including two Single Access (SA) antennas and one phased array Multiple Access (MA) antenna. These two types of ILAs are heterogeneous in operation frequency, information rate, and service type [3]. With the rapid increase of space information generated in recent years, the contradiction between limited ILA and fast-growing service demands is highlighted. So, high efficient scheduling of ILA to complete more services is the key to improve system's performance [4]. Furthermore, because ILA can not provide service during the period of preparation, setup time between adjacent services must be seriously analyzed in the actual scheduling problem. The characteristic of setup time for two types of ILAs is different. MA antennas use electro-sweep mode to point to the user, whose setup time is very short and basically constant, while SA antenna adopts mechanical rotation mode and its setup time is related to the space position of adjacent services and the rotation speed of antenna. We can conclude that the setup time of SA antenna between adjacent services is dynamic and is a principle factor in this paper.

The conventional scheduling models about the TDRSS usually imagine setup time as static and reserve the largest and constant value for this parameter [5–8], which result in serious resource loss of ILAs. In [5,6], a parallel machine scheduling model with time windows is built for the TDRSS. In his work, the tightness of service's visibility window is introduced and a greedy randomized adaptive search algorithm (GRASP) is implemented to solve the scheduling model. Although the setup time factor is incorporated into his model, its dynamic property is not analyzed. In [9], a CSP model considers factors like service priority and scheduling criterion. In [10], the problem of hybrid link scheduling of laser and microwave is studied, and a CSP model for maximizing the sum of the weight (or priority) of scheduled service is established. Neither [9] nor [10] considers the setup time.

To solve the above problem, this paper analyzes the scheduling principle of the TDRSS with dynamic setup time, and translates this problem into the pointing route problem of multiple types of ILAs. As the setup time variable depends on the sequence and start time of services, we introduce a new dynamic constraint into the conventional scheduling model and propose a mixed integer programming model for the TDRSS. For solving this model, we implement one improved algorithm based on the GRASP scheme. In the typical TDRSS scenario, we carry out comprehensive simulations with different data types and problem scales. Numerical results show that, compared with the conventional scheduling method that only considers static setup time, the proposed model and algorithm can effectively improve performance of the TDRSS.

2   Mathematical Formulation

2.1   Problem Analysis

The relay service has the following characteristics: (1) there are multiple types of antennas to provide services [3]; (2) relay process must be within user's visibility window; (3) highly dynamic users usually cover large time-space span during service; (4) as user has been tracking by the rotation of SA antenna or the digital beam forming of MA antenna during relay, it is more suitable to use the pointing angle of ILA to represent the space position of service. In the traditional ground-based routing problem, the spatial relationship between services is analyzed by the distance in the geodetic coordinate system and each service's time-space span is small [11,12]. So, conventional methods on routing problems [13,14] are not fully applicable to our problem.

We first convert the space position of service in the geodetic coordinate system into the azimuth and elevation angle in the antenna coordinate system, and then establish pointing route for each ILA. Last, each pointing route of antenna is projected onto the time line to form a scheduling sequence. Figure 2 shows the scheduling principle of the TDRSS based on dynamic setup times.

In Fig. 2, the vertex 0 corresponds to the start and end point of the pointing route, and the vertexes from 1 to 5 correspond to relay services. The pointing route during relay process and service preparation is represented by a solid line and a dotted line respectively. The relay process must be within user's visibility window and the next service's start time must be greater than addition of the previous service's end time and setup times.

2.2   Mathematical Model

The TDRSS scheduling problem can be defined on a directed graph $G = (V, A)$ in the ILA coordinate system, where V is a vertex set and $A = \{(i,j) : i,j \in V, i \neq j\}$ is an arc set. The graph has $|N| = n + 2$ vertexes, in which node 0 and node $n + 1$ represent the start and end vertex of the pointing route of ILA and other vertexes are services denoted by $N = V \backslash \{0, n + 1\}$. The position of each vertex is represented by the azimuth and elevation in the coordinate system of ILA. For vertex 0 and $n + 1$, both the azimuth and elevation both are zero. All feasible ILA pointing routes correspond to source-to-sink elementary paths in G.

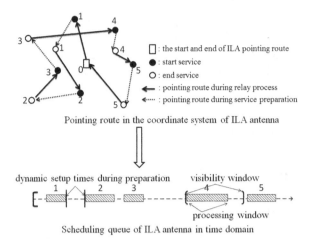

Fig. 2. The scheduling principle of the TDRSS based on dynamic setup times

In order to formulate our problem, the following variables and parameters are defined.

Let K be the set of all heterogeneous ILAs, and be partitioned into subsets of homogeneous ILA $K = K_1 \cup K_2 \cup \ldots \cup K_P$. In our problem, there are two types of ILAs, namely, SA and MA denoted by K_1 and K_2. v_k is the angular velocity of SA antenna rotation during setup time and C_k represents the constant setup time of MA antenna.

Vertex related parameters and variables are defined as follows. Let p_i be the processing time of vertex i processed by ILA k and set $p_0^k = p_{n+1}^k = 0$. Vertex i should be processed within its visibility window denoted by $[w_i^s, w_i^e]$. And particularly for vertex 0 and vertex $n + 1$, there is also a time window $[0, T]$, where T is the scheduling horizon. At any time t in $[w_i^s, w_i^e]$, the position of vertex i is identified by the azimuth and elevation of ILA k, i.e., $(\alpha_{i,t}^k, \beta_{i,t}^k)$. Let $\delta^+(i) = \{j : (i,j) \in A\}$ and $\delta^-(j) = \{i : (i,j) \in A\}$. The setup time from the end position of vertex i to the start position of vertex j at instant t is represented by $s_{i,j,t}^k$.

The formulation also requires two groups of decision variables. The first group models the vertex processing sequence on ILA k, defined by an binary variable $x_{i,j}^k$. If vertex i precedes j on ILA k, $x_{i,j}^k = 1$; otherwise, $x_{i,j}^k = 0$. The second group contains the start time t_i^s and the end time t_i^e of vertex i.

The TDRSS scheduling problem can be formulated as follows:

$$\max \sum_{k \in K} \sum_{(i,j) \in A} x_{i,j}^k \tag{1}$$

subject to:

$$\sum_{k \in K} \sum_{j \in \delta^+(i)} x_{i,j}^k \leq 1, \quad i \in N \tag{2}$$

$$\sum_{j \in \delta^+(0)} x_{0,j}^k = 1, \quad k \in K \tag{3}$$

$$\sum_{i \in \delta^-(j)} x_{i,j}^k - \sum_{i \in \delta^+(j)} x_{j,i}^k = 0, \quad k \in K, j \in N \tag{4}$$

$$\sum_{i \in \delta^-(n+1)} x_{i,n+1}^k = 1, \quad k \in K \tag{5}$$

$$s_{i,j,t_j^s} = \frac{\max(|\alpha_{j,t_j^s}^k - \alpha_{i,t_i^e}^k|, |\beta_{j,t_j^s}^k - \beta_{i,t_i^e}^k|)}{v_k},$$
$$k \in K_1, (i,j) \in A \tag{6}$$

$$x_{i,j}^k (t_i^s + p_i^k + s_{i,j,t_j^s}^k - t_j^s) \leq 0, \quad k \in K, (i,j) \in A \tag{7}$$

$$w_i^s \leq t_i^s \leq w_i^e - p_i^k, \quad k \in K, i \in V \tag{8}$$

$$x_{i,j}^k \in 0, 1, \quad k \in K, (i,j) \in A \tag{9}$$

The scheduling goal is to maximize the number of service completions as shown in (1). Constraint (2) ensures that each service is executed at most once by one antenna. Constraint (3) ensures that each antenna is allocable and starts pointing from zero to the first user. Constraint (4) ensures that each antenna processes a maximum of one service at a time. Constraint (5) requires each antenna to return to zero position after processing services. (6) constrains the dynamic setup time of SA antenna between adjacent services. Constraint (7) represents the time sequence between adjacent services. Constraint (8) ensures that the service is processed in its visibility time window. Finally, (9) limits the value of service flow variables.

3   Algorithm

The TDRSS scheduling problem is NP-hard, which means that it is not applicable for large problem scale to obtain the optimal solution by exact mathematical method [6] in acceptable time. GRASP [15], as an heuristic iterative method, has been widely used in many combinatorial optimization problems. This method can obtain a satisfactory suboptimal solution by a finite number of iterations in reasonable times. We implement a meta-heuristic algorithm based on the GRASP framework that selects the best solution produced by several iterations as the output. Each iteration in our algorithm consists of two phases, including initial solution construction and local search. In the initial solution construction, a candidate service is randomly selected from the Restricted Candidate List (RCL) and inserted into the current ILA pointing route. Phase 1 will terminate until all antennas can not be inserted any more. Then the local search phase starts to search for the optimal solution in the neighborhood structure of initial solution. Compared with the conventional GRASP algorithm in [6], this paper makes the following improvements: (1) we adds the function of iterative computing dynamic setup times and service start time; (2) we further improve the quality of solution by adjusting the length of RCL adaptively based on results of previous iterations; (3) our neighborhood structure in the second phase is simplified and constrained among different inter-route relocations to reduce the search space. The algorithm flow is shown in Fig. 3.

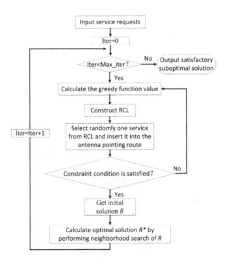

Fig. 3. Flow chart of the algorithm procedure

In the process of constructing RCL, the conventional GRASP fixes the length of RCL and the search space is seriously limited and algorithm robustness is not good enough. Furthermore, as the adjacent iterations are independent of each other, the current iteration can not effectively make use of the information generated by previous iterations. So, in this paper we design an iterative control mechanism for the length of RCL. We choose the optimal RCL length from the candidate lengths by utilizing the solution information generated by previous iterations. The concrete method is as follows.

Define l as the length of RCL, and the candidate length set is $L = l_1, l_2, ..., l_m$. In the first iteration of the procedure, the probability of each element in L is the same, that is, the probability of selection is $p_g = 1/m, g = 1, 2, ..., m$. In subsequent iterations, let S^* be the best feasible solution at present and S_g denotes the average feasible solution generated by all previous iterations. So, before the start of the current iteration, the candidate probability for each element in L is:

$$p_g = \frac{q_g}{\sum_{h=1}^{m} q_h} \tag{10}$$

In (10), $q_g = S_g/S^*, g = 1, 2, ..., m$. If the RCL length $l = l_g$ yields the best feasible solution in the mean sense, q_g will increase and its candidate probability p_g of the length l_g will increase accordingly. Thus, the quality of subsequent feasible solutions can be improved by iteratively using previous solution information.

4 Numerical Analysis

Our scheduling simulation is in the typical operational scenario of the TDRSS as shown in Fig. 1. Each relay satellite has two SA antennas and one MA antenna. So, in our scheduling there is a total of six antennas to provide relay services. In [6], the problem scale is 400, and the total service demand time is close to the maximum time resource that all antennas can provide. In order to analyze our problem under different problem scales, we also consider the service scale 200 (the total demand time is about 50% of the maximum time resource that the antennas can provide) and 600 (the total demand time is about 150% of the maximum time resource that the antennas can provide). The planning horizon corresponds to 86400 s. In each problem scale, five data types are generated according to reference [6]: (1) spltw; (2) spttw; (3) lpltw; (4) lpttw; (5) rand. These data types differ in distributions of service duration time and time window's tightness. Five instances are generated randomly in accordance with each data type, and a total of 75 instances are created for our scheduling analysis. The conventional method (based on static setup times) and our proposed method (based on dynamic setup times) are used to schedule the above mentioned instances with 5 data types and 3 problem sizes. Note that the Table 1 lists the average value of scheduling results belonging to the same data type.

From Table 1, we can see that our method not only improves the total scheduled number (TSN), but also reduces total setup times (TST) of all ILAs and setup times of SA antennas (SAST)for all instances. Due to the addition of calculating the dynamic setup times and the service start time iteratively, our method consumes more CPU times than the conventional.

In order to further analyze the results, we introduce two definitions, namely, scheduling success rate (SSR) and invalid resource consumption rate of SA antenna (IRCRSAA) to describe the satisfaction degree of users' demands and the utilization efficiency of SA antennas.

Specifically, SSR represents the ratio of the number of scheduled services to the total number of service requirements:

$$SSR = \frac{\sum_{k \in K} \sum_{(i,j) \in A} x_{i,j}^k}{|N|} \tag{11}$$

The variables and parameters in (11) are defined in Sect. 2.2. IRCRSAA represents the ratio of accumulated setup times to the total utilizable times of SA antennas:

$$IRCRSAA = \frac{\sum_{k \in K_1} \sum_{(i,j) \in A} x_{i,j}^k s_{i,j,t_j^s}^k}{|K_1| \times T} \tag{12}$$

Table 1. Scheduling results of multiple types of data sets under different problem scales

Data type	Size	The conventional method				Our method			
		TSN	TST	SAST	CPU times	TSN	TST	SAST	CPU times
spltw	200	197	70282	69300	19.6	199	25994	25072	395.1
spttw	200	174	63307	62475	11.1	183	25914	25119	249.5
lpltw	200	198	70552	69562	20.7	199	25359	24467	411.3
lpttw	200	168	60405	59587	9.8	174	23759	22979	226.8
rand	200	180	64672	63787	14.5	185	24287	23410	312.6
spltw	400	313	103387	101587	94.0	351	42800	41045	2458.9
spttw	400	246	81465	80062	34.5	281	38020	36639	1019.2
lpltw	400	282	92970	91350	83.7	316	38013	36430	2158.7
lpttw	400	222	73462	72187	25.3	252	33284	32017	755.9
rand	400	263	88357	86887	56.8	300	38874	37389	1571.9
spltw	600	339	106065	103950	171.2	393	45609	43487	5081.9
spttw	600	267	87142	85575	40.6	310	38775	37207	1404.5
lpltw	600	297	96510	94762	141.2	337	40753	39006	3947.3
lpttw	600	234	76950	75600	29.3	272	34759	33393	961.5
rand	600	286	94560	92925	86.5	336	42526	40884	2789.2

In (12) T is the planning horizon, and the remaining variables and parameters are described in Sect. 2.2.

SSR and IRCRSAA are shown in Figs. 4 and 5, respectively.

It can be seen from Figs. 4 and 5 that, in the scheduling problem of the TDRSS, SSR and IRCRSAA are related to the problem scale and data type regardless of whether or not the dynamic setup times being considered.

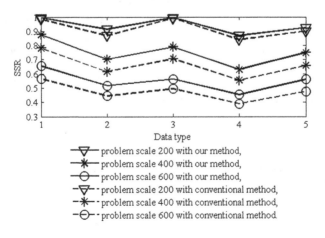

Fig. 4. Scheduling success rate

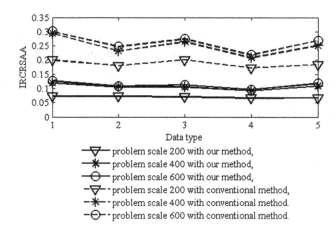

Fig. 5. Invalid resource consumption rate of SA antenna

In Fig. 4, when the problem scale is small (200), the maximum utilizable time that all ILAs can provide is about 2 times of the total demand. The SSR in each methods is more than 90% and our method is 2.3% higher than the conventional. With the increase of problem scale (400 and 600) (the total demand time is getting close to and exceeding the maximum utilizable time that all ILAs can provide), both methods get lower SSR, and our method is 8.7% and 7.5% higher than the conventional for these two scales respectively. We can also see that, for the same scale, the SSR declines in the order of spltw, lpltw, rand, spttw and lpttw. There are two reasons. First, the service with shorter duration time and larger sliding range is more likely to be scheduled. Second, the service sliding range in its visibility window has more significance on the SSR than the duration time factor.

In Fig. 5, with the increase of problem scale from 200 to 400, the IRCRSAA in two methods both increase evidently. When the problem scale continues to increase from 400 to 600, the IRCRSAA increase slowly. For the problem scale 200, 400 and 600, the IRCRSAA in our method is 11.8%, 14.4% and 15.0% lower than that in the conventional.

5   Conclusion

Efficient scheduling is critical for the space-based TDRSS. Taking into account two types of ILAs and actual daily operation with dynamic setup time, we transform the original scheduling problem of the system into the pointing route problem of ILAs and formulate this routing problem with an mixed integer programming model. Then, an improved GRASP algorithm is implemented to solve this model. In the typical operation scenario of the TDRSS, 75 service instances, including 3 problem scales and 5 data types, are designed to evaluate our model and algorithm. Numerical results show that, compared with the conventional

GRASP based on static setup times, our method can reduce the invalid resource consumption rate of SA antennas by 11.8%, 14.4% and 15.0% for the problem scale 200, 400 and 600 respectively, which effectively improves the efficiency of the TDRSS. In addition, as the TDRSS' planning horizon is much longer than the CPU times of scheduling, the CPU time consumed by calculating dynamic setup times is acceptable. In future work, based on the scheduling model in this paper, we will further optimize the algorithm by combining with exact branch-and-price method.

Acknowledgments. This work is partially supported by the National Natural Science Foundation of China (91438206, 91338108).

References

1. Gramling, J.J., Chrissotimos, N.G.: Three generations of NASAs tracking and data relay satellite system. In: 2008 AIAA SpaceOps Conference, pp. 1–11. AIAA press, Heidelberg (2008)
2. Wang, J.S., Qi, X.: China's data relay satellite system served for manned spaceflight. Sci. Sin. Tech. **44**(3), 235–242 (2014)
3. Goddard Space Flight Center/Exploration and Space Communications Projects Division: Space network handbook. Greenbelt, Maryland (2007)
4. Huang, H.M.: Reflections on development of the ground system of the first generation CTDRSS. J. Spacecraft TT&C Technol. **31**(5), 1–5 (2012)
5. Rojanasoonthon, S., Bard, J., Reddy, S.: Algorithms for parallel machine scheduling: a case study of the tracking and data relay satellite system. J. Oper. Res. Soc. **54**(8), 806–821 (2003)
6. Rojanasoonthon, S., Bard, J.: A grasp for parallel machine scheduling with time windows. INFORMS J. Comput. **17**(1), 32–51 (2005)
7. Lin, P., Kuang, L.L., Chen, X., Yan, J., Lu, J.H.: Adaptive subsequence adjustment with evolutionary asymmetric path relinking for TDRSS scheduling. J. Syst. Eng. Electron. **25**(5), 800–810 (2014)
8. Lin, P., Yan, J., Fei, L.G.: Multi-satellite and multi-antenna TDRSS dynamic scheduling method. J. Tsinghua Univ. (Sci. Technol.) **55**(5), 491–496 (2015)
9. Fang, Y.S., Chen, Y.W.: Constraint programming model of TDRSS single access link scheduling problem. In: 5th International Conference on Machine Learning and Cybernetics, pp. 948–951. IEEE Press, Dalian (2006)
10. Zhao, W.H., Zhao, J., Li, Y.J.: Resources scheduling for data relay satellite with microwave andoptical hybrid links based on improved niche genetic algorithm. Int. J. Light Electron Opt. **125**(13), 3370–3375 (2014)
11. Laporte, G.: Scheduling issues in vehicle routing. Ann. Oper. Res. **236**(2), 463–474 (2016)
12. Toth, P., Vigo, D.: Vehicle Routing Problems, Methods and Applications, 2nd edn. Society for Industrial and Applied Mathematics, Philadelphia (2014)
13. Laporte, G.: The vehicle routing problem: an overview of exact and approximate algorithms. Eur. J. Oper. Res **59**(3), 345–358 (1992)
14. Braysy, O., Gendreau, M.: Vehicle routing problem with time windows, part I: route construction and local search algorithms. Transp. Sci. **39**(1), 104–118 (2005)
15. Feo, T.A., Resende, M.C.: Greedy randomized adaptive search procedures. J. Glob. Optim. **6**, 109–133 (1995)

Research on Spatial Information Network System Construction and Validation Technology

Hongbin Zhou[1(✉)], Jingchao Wang[2], and Jincan Liu[1]

[1] The 54th Research Institute of CETC, Shijiazhuang, Hebei, China
frenhz@sina.com, 13810397793@163.com
[2] Institute of China Electronic System Engineering Corporation, Beijing, China

Abstract. Reviews of foreign spatial information network system design and demonstration technology development are introduced on this paper. Spatial information network system model is carried out. Considering about the complex architecture and characteristics, System construction and validation methods, architecture and assessment criterion are researched and integrated supporting environment is designed. It supplied the procedure and validation methods about concept system, technologies and evaluation. Also, the demonstration of its application is supposed here.

Keywords: Spatial information network · Architecture design · System construction · Integrated supporting environment · Performance assessment

1 Preface

At the present stage, the research of the complicated system usually carries on the concept and technology verification in the system demonstration and the demand analysis stage, in order to evaluate the system technical feasibility and maturity. For the spatial information network, the research and construction of system are interactive and developed alternately with the system verification. The system design of spatial information network, guides to the system verification research on the application model, system model, interface model that reflect service characteristic, through reasonable and effective test method, to analyze and verify the function and performance of each constitute elements and their interaction relationship. On the other hand, the achievements of verification support the top-level system design of spatial information network and technical system demonstration, and lead a subsequent new technical research. The spatial information network system construction and validation methods are discussed and analyzed in this paper, and the corresponding solution methods and design ideas of the experimental framework design, experimental methods, evaluation mechanism etc. are proposed.

© Springer Nature Singapore Pte Ltd. 2017
Q. Yu (Ed.): SINC 2016, CCIS 688, pp. 45–53, 2017.
DOI: 10.1007/978-981-10-4403-8_5

2 Research Situation at Domestic and Abroad

In view of the complex system construction and the validation method, many explorations have been developed at home and abroad [1, 2]. Take the United States as an example, United States Department of Defense develops the integrated C4ISR system by using 3 sides/perspectives to describe the system architecture, and finally produces a comprehensive system framework that including combat system, system architecture, technical system. At the follow-up stage, the U.S. put forward C5ISR (C4ISR + Combat) as a new concept, which unifies cognitive domain, information domain and physical domain theoretically, and merges information system and weapon platform into an integral whole. In the other hand, as the developing of system simulation and testing technology, the experimental verification method of the large-scale system is gradually turning from traditional simulation and semi physical simulation method to the system validation methods that based on live, virtual, construction (LVC) structure. On this basis, many distributed environment that permit specific application fields have been constructed, such as GIG-EF, FCS SoSIL, JMETC, NCR, etc., which support cross-domain system demonstration and new technology validation.

In the domestic, the comprehensive technological integration and demonstration in higher level are based on prior research, and have been widely used in recent years in the related research domain, particularly in satellite communications, spatial information network and other research projects. Since 90's, the national medium-and long-term development plan, NSFC, 863 Program have been set up key research topics related to spatial information network, including system architecture, protocol, management, security etc [3–5]. In 2006, Academician Shen Rong-jun put forward Chinese Integrated Space-Ground Network at the first time. Concerning of the rapid development and problems of China's space mission, He proposed an interconnection system integrating inter-satellite links, satellite-ground links, and terrestrial network. This system also integrates the resources of data receiving, transmission, distribution and operation control, that related with spacecraft. The system finally achieves information sharing and coordination construction by offering a variety of information to various terminals. The recently started Space Station Project aims to validate the system architecture, protocol, networking mode, typical applications of the Integrated Space-Air-Ground Network. Compared with the abroad, our country in the research of spatial information network has developed an extensive work and achieved some results in system architecture and key technology, but the researches in the integrated system design, key technology verification and organization application are still insufficient.

3 System Design Method

The spatial information network has huge scale, complex structure and its spatial nodes have high dynamic motion. Besides, the business requirements are changed by services. To carry out the demonstration and verification of such a complex and huge system, a system model which can simulate the network dynamically must be established by scientific method and theory. This model not only reflects the effectiveness of the system, but also validates the components' connection, coordination, and effect to the system by constructing an equivalent model. Drawing on the experience of American C4ISR [6], the spatial information network architecture can divided into three parts: application system, functional system and technical system. To be more specific, the application system reflects the network's capacity; the functional system indicates the network's design indicators; the technical system provides technical standards. The three systems restrict and rely on each other. The validation of the three systems needs some models such as concept demonstration model, system architecture model, performance evaluation model, technical architecture model, interface function model and so on. The research process of the spatial information network is as shown in Fig. 1.

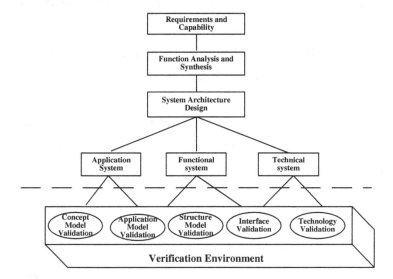

Fig. 1. Research process of spatial information network

4   System Model Construction

4.1   Application System Modeling

In the phase of system demonstration, architecture design should adopt an advanced system verification tool, and structure the system model scientifically. By validating the system architecture design, we can verify that the architecture design can meets the functional requirement, and the system constructed by the design can meets the application requirement. Generally speaking, the system modeling includes several steps: first to analyze the system demand and create the system use cases; secondly, analyze the system function and the correlation between components; at last, decompose the system and construct each of the subsystem. In this way, the system modeling is accomplished. Then the function system and the technical can be designed according to the system model.

Considering the system requirements, the spatial information network takes the space platform (e.g. GEO, MEO, LEO, near space platform and aircraft etc.) as the carrier to acquire, transmit and process spatial information in real time. As an important national infrastructure, the spatial information network can satisfy many major applications, such as ocean-going voyage, emergency rescue, air transportation, space TT&C, navigation and positioning. Besides, it supports high dynamic and real-time broadband transmission in the earth observation; it tolerates ultra-long range and large time-delay in the deep space exploration.

According to the above-mentioned requirements, we establish the application system for spatial information network by using Unified Platform for Defense Modeling (UPDM). Figure 2 describes the system's key components and the logical correlation between them, and the detailed description as follows.

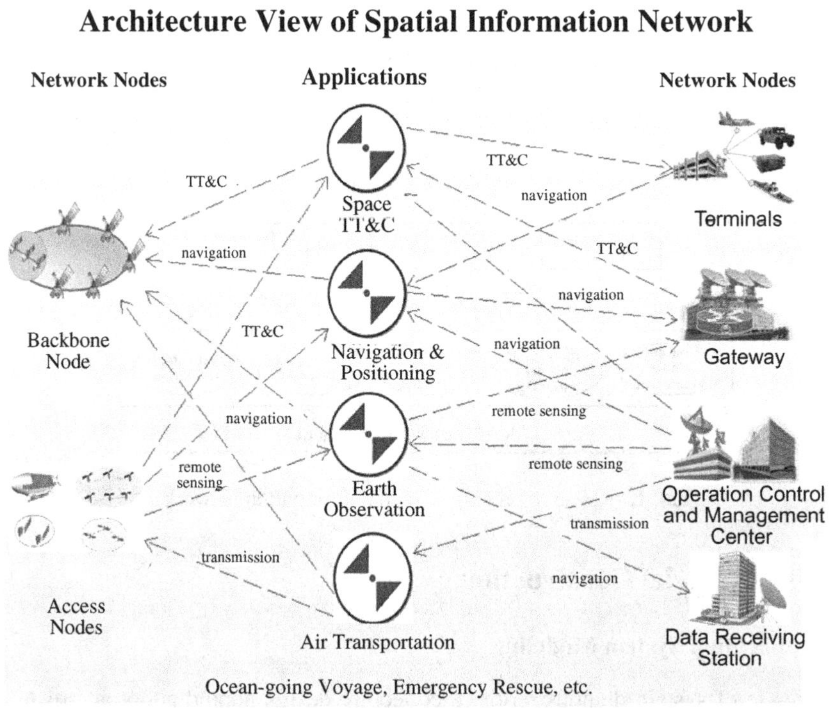

Architecture View of Spatial Information Network

Fig. 2. Application system of spatial information network

The nodes in space segment consist of the multi-functional satellites that distribute in GEO/MEO/LEO and other space-based platforms. They are layered deployed and they acquire a variety of information including information perception, intelligence distribution, command and control, etc. Their mission is providing broadband transmission, routing switching, information processing and so on.

In ground segment, the network nodes are responsible for supporting the network's operation, including all of the infrastructure that relate to the nodes in space segment, such as the Operation Control and Management Center, the Gateway Net, the Data Receiving Net, the Ground TT&C Net, etc. In particular, the Operation Control and Management Center is responsible for task planning, resources allocation, network management and control; the Gateway Net realizes the information access, switching, transmission, and achieve intercommunication with the ground network; the Data Receiving Net collects all kinds of raw data from the nodes in space segment, accomplishing integration of the information processing, information fusion and product generation; the Ground TT&C Net is in charge of the aircraft's TT&C, data delay, data synchronization and so on.

The customers include various kinds of application satellites, user spacecraft, user vehicle, and all of the terminals distributed the whole world.

In Fig. 2, the links indicate the logical relation between the nodes about different application.

4.2 Function System Modeling

In the system function analysis stage, the main work is capability analysis and integration, according to the functional requirement that received from the application system model. The key point is transforming the functional requirement to system capability, and the top-down modeling method should be adopted. After confirming and distributing the required function, the components acquire their subsystem, state transition diagrams and the interfaces. Validating the subsystem model, we can verify the response in the specific state when the system is working.

Using the design method mentioned above, the functional system and of the components as shown in Fig. 3.

In the Fig. 3, the backbone node is the network platform for information aggregation, relay switching and service convergence, it includes the on-board data process module, the route switching module, the network control module, realizing the information acquisition, processing, transmission and aggregation. The backbone nodes compose a space-air integrated network by inter-satellite-links, satellite-to-aircraft links and air-to-air links. The gateway includes access control module, route switching module, resource reservation module, protocol conversation module, etc. The gateway has links to the backbone nodes, the access nodes and the ground network, in order to realize the intercommunication with the ground network. The terminals contain access module, routing module, wireless transmission module, etc. The terminals' capabilities include network access, safety certification, resource application, data packaging, etc. Based on the functional system, the technical system can be planned and designed.

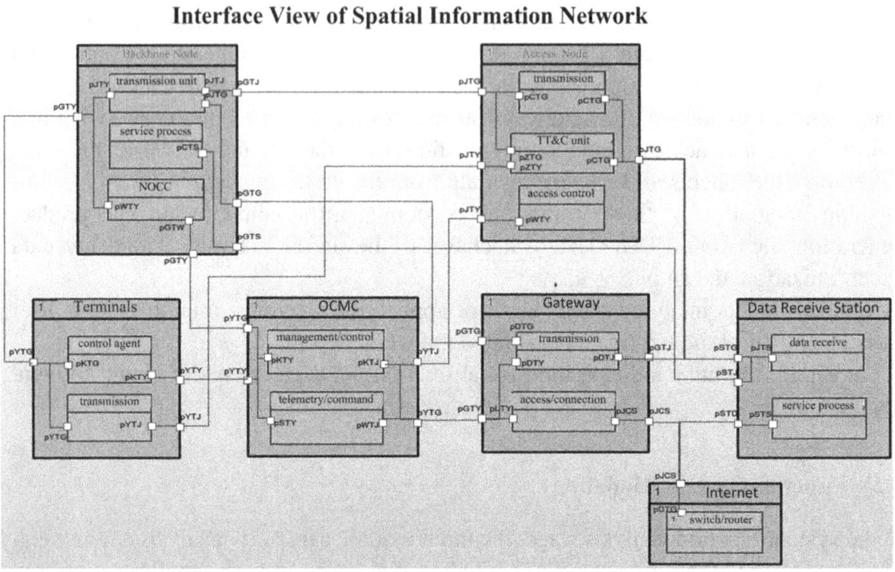

Interface View of Spatial Information Network

Fig. 3. Functional system of spatial information network

5   System Validation

5.1   Integrated Supporting Environment

The verification of spatial information network should contain validate the function deployment, the correlation between subsystems, the work flow, etc. Therefore, the supporting environment comprises the architecture modeling, orbit design, network simulation and task demonstration. Based on this, the environment integrates the architecture, technology and application verification.

As mentioned, the environment consisted of UPDM, STK, OPNET as the universal simulation platform and the physical experiment system. On top of this, the environment provides varieties kinds of model base, such as architecture design model, network simulation model, orbit model, protocol model, link model, etc. Besides, the environment has the interface to the physical experiment system, and obtains the experimental data to support application and technical system validation.

As shown in Fig. 4, the system architecture modeling is based on UPDM, orienting the combat mission. By using formal description, UPDM construct the architecture model and work flow to satisfy the system requirements [7]. During the task demonstration, we can verify the system architecture, in other words, UPDM provides the validation in the system design stage. Constellation configuration and orbit simulation are based on STK, and designed together with the network simulation. STK verify the constellation design, covering capacity, satellite orbit, topological stability, etc. The results of simulation are output to the network simulation. The validation of technical system is achieved by network simulation or the physical experiment system, verifying the system function and key technology, such as integrated space-ground routing ability, access control ability, QoS, etc. Finally, for the different application verification, the environment validates the system application and task flow with the help of STK 3D demonstration. The display of integrated network is real-time, and the evolution process is dynamic.

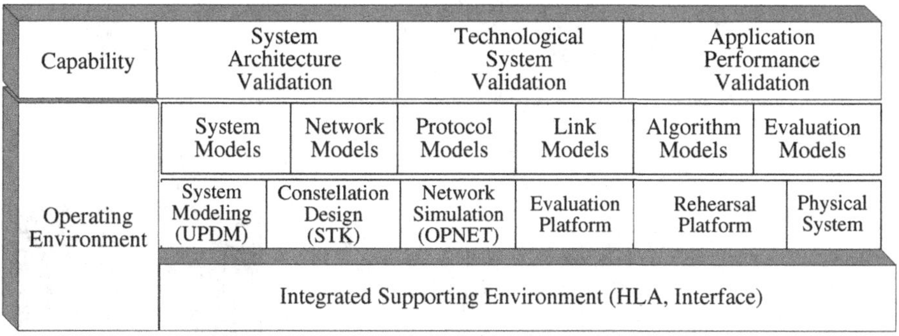

Capability	System Architecture Validation		Technological System Validation		Application Performance Validation	
	System Models	Network Models	Protocol Models	Link Models	Algorithm Models	Evaluation Models
Operating Environment	System Modeling (UPDM)	Constellation Design (STK)	Network Simulation (OPNET)	Evaluation Platform	Rehearsal Platform	Physical System
	Integrated Supporting Environment (HLA, Interface)					

Fig. 4. Integrated supporting environment for system validation

5.2 Application Example

In order to accomplishing the top-layer designing of spatial information network, this paper proposes the system architecture modeling, orbit design, network simulation and task demonstration, that based on the system application. The verification environment mentioned above, enable the concept system validation, technology verification and application evaluation. An example is shown in Fig. 5.

Step one: The system construction and architecture verification.

As described in Sect. 4, the spatial information network function system, information flow and system architecture model can be built on the platform of system modeling, which based on the system requirements.

To verify the architecture model, a sequence diagram should be constructed based on the overall concept plan or the mission scenario in the supporting environment. During the model validation process, the sequence diagram would be triggered by a typical event in the architecture model, and go through the state transition diagrams according to the external parameters and trigger condition. In this way, the function of network nodes are tested, and the mission concept, interfaces, work flow, typical application of the system are verified.

Step two: The deployments in space segment and constellation design.

According to system architecture model, the deployments in space segment meet the system requirements. The constellation design and network simulation are combined design using STK and OPNET. The verification of constellation design includes covering capacity, satellite orbit, topological stability, etc. and the results of simulation are output to the network simulation and mission rehearsal.

Step three: The technical system determination and validation.

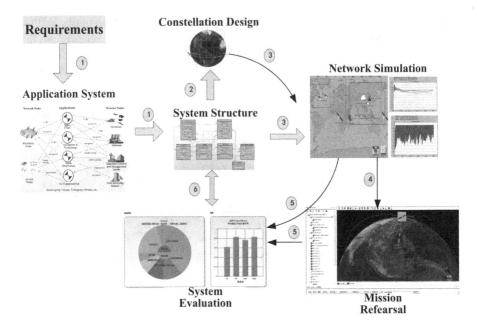

Fig. 5. Application example

Based on the architecture design, the verification of functional system and technical system are necessary. The validation contents include network/link communication capability, network control, interconnection performance, quality of service for different needs, etc., and support the technological optimization. All of the simulation result and experimental data are reliable source for system evaluation and demonstration.

Step four: The validation of typical application and mission rehearsal.

According to the system application and service model, several mission rehearsals should be constructed to verify the assurance ability of combat mission. During the demonstration, the state of integrated network and dynamic evolution process is displayed.

Step five: The evaluation of comprehensive performance.

Constructing the index system and performance evaluation model is the basis of system assessment. For typical applications, the evaluation model analyzes the simulation results and experimental data, and achieves the comprehensive evaluation, validating if the architecture design meet the system requirements or not.

The spatial information network needs to be verified the logical correctness of architecture design, technological feasibility and satisfaction degree for application requirements, therefor it is important to construct comprehensive performance evaluation system, which analyzes and synthesize the function of system components. Besides, the selection and quantification of indicators is the key. During the evaluation modeling, the relationship between components and the effects on the system should be analyzed to construct evaluation model and verify the performance with multi angle.

Step six: Optimize the architecture design based on evaluation results.

6 Conclusion

It is necessary to adopt scientific methods and validation means for spatial information network to carry out the verification of system construction and technical feasibility. This paper researches on experimental architecture framework, and proposes a comprehensive verification environment for architecture modeling, orbit design, network simulation and mission rehearsal, based on the application of new methods, new technologies and the ability to accurately assess the system. The design flow and the verification of concept system, technological system and application system are proposed, which can provide technical support for spatial information network demonstration, system design, and technology research and so on.

References

1. Gao, B.-N.: Analysis and Design of Demonstration Experiment Management System for Information Systems. Beijing University of Posts and Telecommunications, Beijing (2011)
2. Huang, Z.-X.: Research on system integrated and validation technology of C4ISR. Silicon **15**, 35–36 (2012)
3. Shen, R.-J.: Some thoughts of Chinese integrated space-ground network system. Eng. Sci. **8**, 19–30 (2006)
4. Min, S.-Q.: An idea of China's space-based integrated information network. Spacecraft Eng. **22**(5), 1–14 (2013)
5. Jun, Z.: Space–Based Mobile Communications Network. National Defense Industry Press, Beijing (2011)
6. DoD Architecture Framework Working Group. DoD Architecture Framework. vol. II, Product Description Version1.5 (2007)
7. Sun, Y., Wang, G.-W.: Weapon equipment system modeling based on UPDM. J. Acad. Armored Force Eng. **26**(2), 66–69 (2013)

Microwave-Laser Integrated Network Component Technology

Xiujun Huang[⊠], Dele Shi, and Zongfeng Ma

Shandong Institute of Aerospace Electronics Technology, Yantai, China
`xiujunmail@163.com, sh_dl@163.com`

Abstract. Distributed reconfigurable satellite is a new kind of spacecraft system, which is based on a flexible platform of modularization and standardization. Based on the module data flow analysis of the spacecraft, this paper proposes a network component of Microwave-Laser integration architecture. Low speed control network with high speed load network of Microwave-Laser communication mode, no mesh network mode, to improve the flexibility of the network. Microwave-Laser integrated network component technology was developed, and carried out the related performance testing and experiment. The results showed that Microwave-Laser integrated network components can meet the demand of future networking between the module of spacecraft.

Keywords: Distributed reconfigurable satellite · Microwave · Laser · Network · Communication · Mesh

1 Introduction

Distributed reconfigurable satellite is the traditional "integration" satellite [1], through the function decomposition, the physical separation of "decoupling physical coupling constraints between systems, through the" formation flight, network connections and system reconstruction and functional reorganization, which can effectively solve the problem of spacecraft application flexibility and anti risk ability, which has the survival strong ability and easy to launch and upgrade technology, manufacturing and low cost [2, 3]. Networking module spacecraft using wireless self-organizing network, each module wireless communication module, support data transmission and routing between modules and module of the spacecraft can be used as any other modular spacecraft communication access point and automatic search module of the spacecraft near to join the cluster [4]. Through the self organizing network, any module can communicate with the ground system through the data transmission module, as shown in Fig. 1 [5].

According to the characteristics of the module group network, we design the network architecture, and put forward the design of Microwave-Laser integration Technology, We combine laser, high speed point-to-point communication and microwave characteristics of low speed, easy to network organically, and develop sample machine of Microwave-Laser integration networks Component Technology and demonstration, the research of this topic, to provide effective support for future applications for distributed reconfigurable satellite system.

© Springer Nature Singapore Pte Ltd. 2017
Q. Yu (Ed.): SINC 2016, CCIS 688, pp. 54–65, 2017.
DOI: 10.1007/978-981-10-4403-8_6

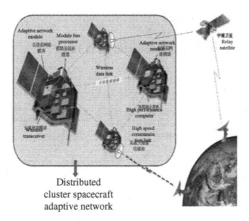

Fig. 1. Schematic diagram of distributed reconfigurable satellite in orbit operation

2 Microwave-Laser Integrated Network Component Architecture

The distributed reconfigurable system mainly contains the following data information which needs to be exchanged: remote information, telemetry information, voice information and image information, payload data, forwarding data of other module spacecraft. As a node of the space communication network, the module stores the data and information transmitted by the satellite link, which is transmitted to the next network node by a certain routing algorithm.

For the module group network, it can be divided into two logical networks: on orbit transportation control data transmission network and load high speed data transmission network. In order to meet the high reliability, network reconfiguration requirements, establish links between modules should be as much as possible, while the use of mesh [6] structure and no center can better meet the requirements, using the omni-directional antenna in order to establish a simple, efficient, reliable and high connectivity connection. The high speed data transmission uses point to multipoint PMP structure [7], which can solve the contradiction between the high speed data transmission power and the low

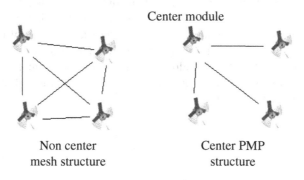

Non center mesh structure	Center PMP structure

Fig. 2. Module group ad hoc Laser networks architecture

power consumption demand of the module, the network component of ad hoc Laser networks architecture is shown in Fig. 2.

Network node design block diagram is shown in Fig. 3.

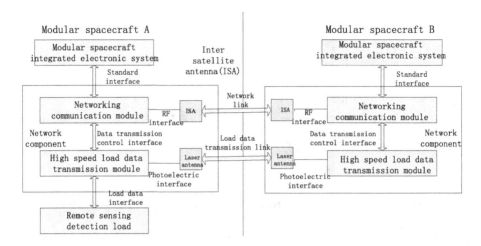

Fig. 3. Network node design block diagram

Network node with the Ad hoc network design of microwave-laser communication mode. For network signaling, environment detection, navigation and control, such as control data, through the microwave communication of control network, complete the complex interactions between modules, at the same time for high-speed data transmission of the optimization of transmission path, through a single jump, jump to deliver data to the final destination. Load high speed data transmission network with the method of laser communication, can realize high speed data transmission, has small volume, light weight and low power consumption, confidentiality and strong anti-jamming capability, and frequency of application, has incomparable superiority in future satellite communications. Mainly includes the following functions:

(1) The network sets up function

Distributed cluster aircraft consists of many modules of the spacecraft, to build into a module of the network, the module can be independent to join, exit, to join the adaptive extension of network scale with the module. After a certain number of orbital period adjustment form initial configuration, and establish the cluster network through information interaction of initial resource allocation and routing information, so you need to cluster aircraft needs to have the function of the network sets up, Module group of ad-hoc network as shown in Fig. 4.

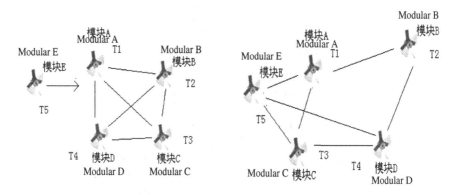

Fig. 4. Module group of ad-hoc network

(2) Network has the ability of organization

Self-organizing network [8] is a kind of adaptive network, in the face of changes in the network environment can be automatic response, have a certain intelligence. Self-organizing main characterization of network organization, number of nodes can be self-organization network formation, and implement the management, including new module registration, time synchronization, routing information exchange and navigation information; In module exits, send out statements, safety evacuation orbit, routing information update and resource redistribution; When allocating new mission requirements, can the path of the complete system configuration is optimized according to the transformation.

(3) Multitasking payload

When after the completion of the module of network building, network can support all kinds of control information interaction between the various modules and data transmission task load.

(4) Network fault diagnosis and recovery

When the cluster module after a failure is detected and designed accurate positioning, isolate the fault module, System independent repair or through the ground remote control to restore failure module, Other modules through independent consultation resources reorganization, thus make the system restore operation.

3   Design of Microwave Communication System

Transport control network measurement and control unit block diagram as shown in Fig. 5, The utility model comprises a radio frequency transmitting module, a radio frequency receiving module, a baseband module and an interface (management unit) module.

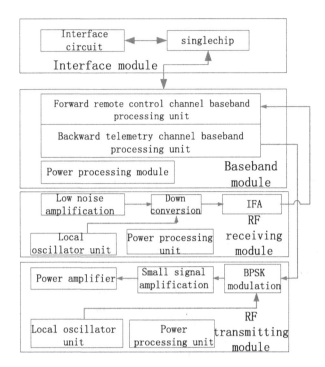

Fig. 5. Transport control network measurement and control unit block diagram

3.1 RF Transceiver Channel Design

The utility model is composed of three parts, namely, the RF module of the transmitting module, the power supply part of the transmitting module and the receiving module. The main functions are as follows:

The radio frequency receiving part is that the S band RF signal received by the antenna is amplified, filtered, and converted to output the intermediate frequency signal of the 66.5 MHz such as to the baseband part of the process;

The RF part of the spread spectrum signal from transmitted baseband. Modulated by BPSK, amplified to antenna.

The RF module has a reference clock for the baseband; complete switch, transmitter module switches and other functions; provide telemetry parameters for the baseband module, baseband acquisition and transmission; at the same time to provide power baseband filter.

(1) Receiving channel design

It consists of five parts: low noise amplifier unit, down conversion unit, if amplifier unit, local oscillator, power supply unit, as shown in Fig. 6. Taking into account the low noise figure of the receiving channel, and the requirements of image rejection and spurious signal suppression, the reasonable components and the gain distribution are selected to meet the requirements of the system noise; Through the design of input frequency selective filter and low noise amplifier with image rejection filter, improve the anti-interference ability of the receiving channel.

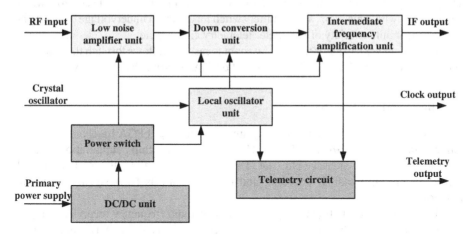

Fig. 6. Receiving channel electrical schematic

In the whole receiving channel, all of the broadband source devices are selected, and the gain fluctuation in the band is less than 0.5 dB. The introduction of the two stage filter in the intermediate frequency amplification unit can guarantee the 3 dB bandwidth and the 40 dB bandwidth of the receiver.

(2) Transmit channel design

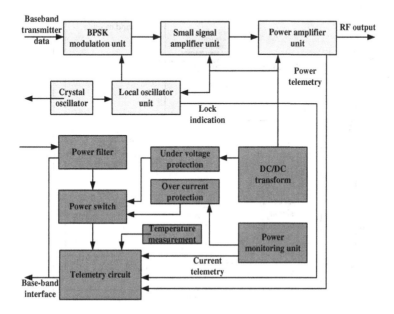

Fig. 7. Transmitter RF channel electrical schematic

The transmitting channel comprises a modulation unit, a small signal amplifying unit, a power amplifying unit, a local oscillator and a power supply, as shown in Fig. 7. The power supply unit comprises a DC-DC conversion circuit, a remote control switch circuit of the whole machine, a transmitter switch machine, an undervoltage overcurrent protection circuit, a variety of parameter telemetry circuits, etc.

The final power capacitor tube fed by shunt measures, It flows through the capacitor current is reduced by half, parallel redundant design, improve the reliability.

3.2 Baseband Module Design

The baseband module includes baseband signal processing, FPGA baseband, configuration management FPGA, ROM/FLASH, ADC and DAC chip configuration, remote control and telemetry interface chip, DC-DC chip, mainly realizes the modulation and demodulation, despreading, encoding and decoding of baseband signal processing function.

The receiving part will output the frequency of radio frequency receiving channel for uplink 7f0 intermediate frequency spread spectrum signal of the A/D sampling into FPGA, fast acquisition by FPGA on the carrier, and the pseudo code tracking and despreading, complete synchronization and data demodulation, the final Viterbi decoding output command data also indicate the working status of the digital baseband; The transmitting part receives the remote input PCM code and the synchronous clock of the external input, and the digital baseband is used for the convolution coding, the

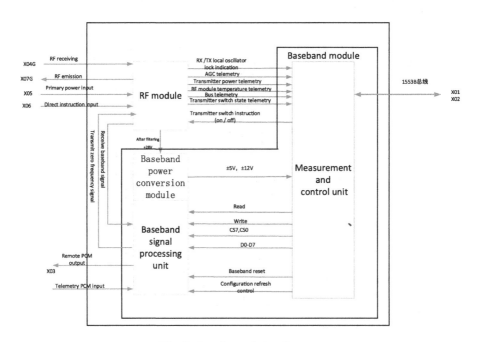

Fig. 8. Interface relation diagram

spread spectrum modulation and the pulse shaping of the telemetry data, and the data and the DA conversion clock are sent to DA processing.

In the downlink, the convolutional coding and the spread spectrum modulation of telemetry data are completed, and the zero if spread signal is generated after the rise of the cosine pulse. After the digital analog conversion, the BPSK modulation is performed in the RF module.

The interface module is realized by MCU, data memory, program memory, AD converter, 1553B bus controller and command driven output circuit. The signal block diagram is shown in Fig. 8.

4 Laser Communication System

High speed load data transmission module based on free space optical communication technology [9–11], A typical space laser communication system is composed of two parts of a transmitter and a receiver, which includes the laser source and the modulation/demodulation unit, optical relay unit, optical send/receive antenna, detection receiver, beam capture, data and signal processing unit and used to perform system working mode and complete signal processing function of main control unit, etc., and its basic function block diagram as shown in Fig. 9.

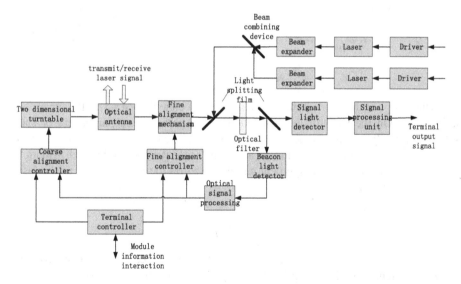

Fig. 9. Space laser communication system functional block diagram

System consists of optical antenna and the stabilizing platform. Optical antenna transceiver communication functions. Stabilized platform realize the APT function. Optical antenna, transmitting channels for the double launch, one for the two receiving channel signal communication, another for four quadrant detector. The working principle of the system:

Modulated signals generated by digital baseband signal modulated by the power drive circuit makes the laser light, so carrying signal through optical laser transmitting antenna. When receiving, at the other end of the laser communication by optical signal receiving antenna will collect to converge at the photoelectric detector and converts the light signal into electrical signal, after amplification, using threshold detection method to detect the useful signal, then restore the baseband signals after demodulation circuit. In the process, APT subsystem implementation on both ends of the laser beam capture, tracking and targeting, optical communication link is established.

Optical antenna is composed of transmitting antenna and receiving antenna. Transmitting antenna is composed of launch fiber and lens, receiving antenna consists of a receiving lens and receiving optical fiber. Transmitting antenna expanded optical output light beam collimation (i.e., compression Angle of light beam) and then launch into the free space。 The receiving antenna to receive light converge into the receiving optical fiber, Complete the function of the optical transmission. The light path principle of optical antenna is shown in Fig. 10.

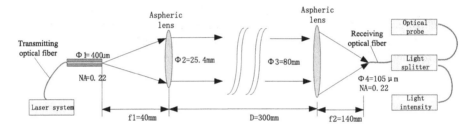

Fig. 10. Laser optical wireless transmission system

In the design of the laser communication system, we should first assign the subsystem level index. On the basis of the realization of the performance parameters of each subsystem, firstly, to ensure that the optical power received by the receiver can meet the requirements of the receiver sensitivity, and then carry out a detailed index distribution.

The basis of the index assignment can be based on the link equation of the laser communication system:

$$P_r = P_t \eta_t \eta_r e^{-\alpha L} (\frac{D_r}{\theta L})^2 \tag{1}$$

In the formula, P_t: the transmitter power; P_r: the power received by the receiver; η_r: receiving antenna efficiency; η_t: Transmitting antenna efficiency; L: Total length for laser transmission; θ: Divergence angle of laser beam; D: Receiving antenna aperture; $\gamma(\lambda)$: Attenuation coefficient of laser transmission in atmospheric channel.

Because of the receiver receives the light power and the square of the launch Angle θ is inversely proportional, emitting angle must be small enough to get enough power receiver. In order to achieve reliable communication, the receiver power must be greater than the sensitivity of the receiver.

P_t: Transmit power is 20dBm, Transmit/receive antenna efficiency is 50%; N_r:Transmit antenna loss is approximately −3 dB; N_t: receive antenna loss is approximately −3 dB; N_a: Atmospheric loss 0 dB; P_{re}: Receiver sensitivity −30 dBm.

The link redundancy can be 14.64 dBm. Laser light wireless transmission system prototype is shown in Fig. 11.

Fig. 11. Laser light wireless transmission system prototype

5   Principles of Prototype Development and Testing

Optical network component principle prototype is shown in Fig. 12, the system uses a centralized management system structure, including CPU module, Wireless ad hoc module, wireless laser communication module and the communication and tracking system channel. The processor is used for on-board processing, wireless ad hoc networks for information transmission platform, The wireless laser communication interface is used to transmit the load information, Plug and play bus interface is used to expand the load. The measurement and control channel is used to realize the platform's satellite ground control. The topology display interface is shown in Fig. 13.

Fig. 12. Optical network component principle prototype

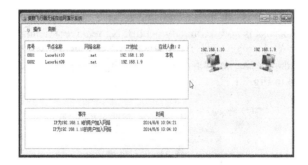

Fig. 13. Topological structure display interface

At the same time, the indoor and outdoor of the Ad hoc network test and laser communication test are carried out, The function of the module's entry and exit is fully verified. The transmission rate test is shown in Table 1.

Table 1. Test result

Link	Data rate Mbps	Transmission direction	Band	Communication system	Networking mode	Antenna direction
Microwave	3	Two-way	S	TDMA/QPSK	Mesh	Omnidirectional
Laser	400	Two-way	1550 nm	PPM	PMP	Directional

6 Conclusions

Distributed reconfigurable satellite is a new kind of spacecraft system, which is a kind of flexible platform based on modular and standardized design. Through the research of this topic, we put forward a kind of Ad hoc laser network component architecture. Low speed control network with high speed load network of Microwave-Laser communication mode, no mesh network mode, to improve the flexibility of the network. Ad hoc Laser networks component technology was developed, and carried out the related performance testing and experiment. The results showed that ad hoc Laser networks components can meet the demand of future networking between the module of spacecraft.

Acknowledgements. Thanks to project team members.

References

1. Atri, D., Panagiotis, T.: Network flow formulation for cooperative peer-to-peer refueling strategies. J. Guid. Control Dyn. **33**(5), 1–11 (2010)
2. Bai, Z.F., Zhang, Y., Shen, L., et al.: Research on rapid access to space of fractionated spacecraft modules.In: Proceedings of the 61th International Astronautical Congress, B4, pp. 5–8. International Astronautical Federation, Prague, Czech Republic (2010)
3. Mathieu, C., Weigel, A.L.: Assessing the flexibility provided by fractionated spacecraft. In: Space 2005 Conference, pp. 1–12. American Institute of Aeronautics and Astronautics, Long Beach, California (2005)
4. Brown, O., Eremenko, P.: Fractionated space architectures: a vision for responsive space. In: Proceedings of the 4th Responsive Space Conference, vol. 1102. American Institute of Aeronautics and Astronautics, Los Angeles, CA, USA (2006)
5. Ma, Z., Xin, M., et al.: Survey on fractionated spacecraft cluster. Spacecr. Eng. **22**(1), 101–105 (2013)
6. Akyildiz, I.F., Wang, X.: Wireless Mesh Networks. Wiley Publications, New York (2009)
7. Babu, V.R., Ghosh, C., Agrawal, D.P.: Enhancing Wireless Mesh Networks Using Cognitive Radios. LAP Lambert Academic Publishing, Fairfield (2011)
8. Li, F., Wang, Y.: Routing in vehicular ad hoe networks: a survey. IEEE Veh. Technol. Mag. **2**(2), 12–22 (2007)
9. Ke, X., Xi, X.: Wireless Laser Communication. BUPT press, Beijing (2004)
10. Cesarone, R.J., Abraham, D.S., Deutsch, L.J.: Prospects for a next generation deep space network. Proc. IEEE **95**(10), 1902–1915 (2007)
11. Kunimori, H., Shoji, Y., Toyoshima, M., Takayama, Y.: Research and development activities on space laser communications in NICT. In: Proceedings of SPIE, vol. 7199, pp. (719904-1)–(719904-7) (2009)

Confer on the Challenges and Evolution of National Defense Space Information Network for Military Application

Yong Jiang[1,3(⊠)], Shanghong Zhao[1], Shihong Zhou[2], Yongjun Li[1],
Yongxing Zheng[1,4], and Haiyan Zhao[1]

[1] Information and Navigation College,
Air Force Engineering University, Xi'an, China
v.jiangyong@hotmail.com
[2] Shanghai Institute of Satellite Engineering, Shanghai, China
[3] 96620 Troops of PLA, Baoding, China
[4] PLA Chongqing Communication College, Chongqing, China

Abstract. As Chinese interests expensing into the oceans, space and electromagnetic areas, single Space-based information systems had been difficult to meet the growing needs of information acquisition, processing and transmission. Building a National Defense Space Information Network by integrating and interweaving space-based systems, was an important and urgent task to construction of information infrastructure for National defense and army, and was also an important guarantee for national security. Firstly, the concept and the connotation of National Defense Space Information Network were analyzed, and then the concept, composition and construction points of network were clearly defined. Secondly, the development course and the research status of American military space-based information systems were introduced in detail, which provided useful reference for China construction of National Defense Space Information Network. Thirdly, the "should be" architecture of National Defense Space Information Network was revalidated, then the key problems of network were analyzed in five aspects including network topology, transmission technology, node processing, and operation-maintenance-management-control technology and security protection. At last, some research advisements were given for China's national defense construction.

Keywords: Military application · National Defense Space Information Network · Hybrid links of laser and RF · Topology control · Onboard switch

1 Introduction

In the current and future, crucial challenges to border security, overseas interests, space competition and other aspects will be faced in China. The intensity and scale of sea territorial controversy, overseas interests maintenance, international peacekeeping tasks, counter-terrorism missions, jointed military exercises, and oceangoing escorts will be increased continuously. The high-tech local war will be the chief battle form, which puts forward higher demand to reliability, mobility, security, compatibility and real time of

© Springer Nature Singapore Pte Ltd. 2017
Q. Yu (Ed.): SINC 2016, CCIS 688, pp. 66–79, 2017.
DOI: 10.1007/978-981-10-4403-8_7

Space-based Information Network. In the angel of information network supporting combat, the shortcomings of current space-based system were exposed, such as lower degree of integration, more difficulty to interconnection, less quick-reaction capability, lower utilization ratio of information resources.

Oriented to PLA's IT-based combat system, military operation other than war and missions of integrated joint operation in the future, to construct a new generation NDSIN (National Defense Space Information Network) with abilities of more tailored, higher security, more convenient, faster responsive, and more comprehensive information supporting, was an important and urgent task to enhance the national defense and army information infrastructure construction.

2 Conception and Connotation

There was none clear definition to space information network. Currently, some concepts were proposed and researched in China academic community. Chinese integrated space-ground network system was proposed by Academician Shen Rong-jun [1] in 2006. It was connected by space and ground links to be an integrated network, emphasizing the integration of space system resources. A special project of "spatial information network" was issued by the National Natural Science Foundation of China [2] in 2013, emphasizing the integration of space-based information carriers such as aircraft, spacecraft and satellites. In 2013, "space integrated information network" was proposed by Min Shi-quan [3]. The basic idea was to realize integration of different satellite systems by communication satellite network. The B&R (the Belt and Road) spatial information corridor program was proposed by NDTIC (National Defence Technology & Industry Committee) [4], the main construction idea was to provide spatial information supporting for economic development, social progress and people's livelihood improvement of countries and regions along the B&R line.

Based on the description of spatial information network above, the NDSIN proposed in this paper was a further refinement from the point, line, and surface description of this concept. And its composition and function was clearly defined as defense space infrastructure with standardization system specification in detail, having a variety of spacecraft as network nodes, with microwave and laser as the main transmission means, having capabilities of global coverage, network information acquisition, storage, processing, distribution, intelligent operation and management. In order to meet the future military application requirements of flat command system and the information joint operations, combined with the military operations, the NDSIN pay more attention to network system security, rapid response capability, rapid deployment capability and saturation information supporting capability.

3 The Development of U.S. Military Space Information Network

In order to fully access and use various information resources to support operation, worldwide military powers leading by U.S. had invested a lot of manpower and resources in the development and construction of the space-based information system.

The construction status of U.S. space-based information system was given in Table 1. From the view of reconnaissance and surveillance, communications, navigation and other typical system, U.S. army had formed a complete Space-based information operations support system.

Table 1. The development of U.S. military spatial information network

Type		Typical system	Function	Technical index	Orbits
Reconnaissance and surveillance	Imaging reconnaissance	KEYHOLE	photoreconnaissance	KH-12 ground resolution 0.1 m,	Perigee 265 km, apogee 650 km
		Lacrosse	Radar imaging	Ground resolution 1–2 m	Perigee 667 km, apogee 692 km, inclination 57
	Electronic reconnaissance	Mentor	Phased array electronic reconnaissance	Scan frequency 100 M–20 GHz	5 GEO
		Trumpet	Electronic reconnaissance	Thousands of signals monitor	3 GEO
		SB WASS	Navy ocean and AF surveillance	Global surveillance double networking	5 × 10 LEO, altitude 1000 km, inclination 63.4
		SBR	Battlefield targets tracking	None	None
	Missile warning	SBIRS	Infrared warning	2.7um & 4.3um sensor	5 GEO + 2 HEO
		DSP	Missile detection	None	3 GEO
		STSS	Ballistic missile detection	Infrared surveillance	2 GEO
	Meteorological	DMSP (7th generation)	Cloud cover, precipitation, ice cover, sea surface wind speed etc.	Ground resolution 300 m, scan 3000 km	5 Sun-synchronous orbit, altitude 830 km
Communication		TDRSS (3rd generation)	Relay	S\C, 800 Mbps; Ku\Ka, 300 Mbps	4 GEO
		MUOS	Narrowband	UHF, 16 beams, capacity 40 Mbps	2 GEO
		WGS	Wideband	Ka & X, 19 beams, single capacity 3.6 Gbps	6 GEO
		AEHF	Protected	EHF & SHF, 37 beams, capacity 430 Mbps	3 GEO
		Iridium	Global	L, 128 Kbps	Altitude 780 km, 66 polar constellation
		APS	Polar communication	EHF	2 GEO
		DSCS	National defense communication	VHF	13 GEO
		GBS	Wideband broadcast	Data, image & video	3 GEO
		MILSTAR	Wartime confidential communication	EHF	5 GEO
		TSAT	Network centric warfare	EHF & Laser	5 GEO
Navigation		GPS	Navigation, position, timing	Positioning accuracy 0.5 m	27 MEO constellation

The TCA (Transformational Communication Architecture) project [5] was proposed by U.S. government in 2002, this project could provide multiple communication services such as high throughput, accessibility, reliability, anti-jamming and anti-interception, and could make the American global combat persons interconnecting through the information network. Shown in Fig. 1, TCA network would compose by WGS (Wideband Global Satellite), MUOS (Mobile User Object System), AEHF (Advanced Especial High Frequency), APS (Advanced Polar System) and TSAT (Transmission Satellite). To change current "chimney" development pattern, TCA project envisaged the various satellite systems of U.S. to achieve effective coordination and networked, could promote information sharing, utilization and integration between systems.

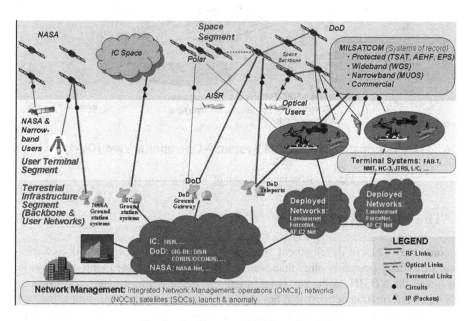

Fig. 1. Architecture of TCA project [5]

Shown in Fig. 2, TSAT [6] was the first step and key point of the TCA project. It was a new generation laser-based and classified wideband spatial information system. Using onboard processing, IP routing and hybrid RF/Laser ISL (Inter Satellite Link) technologies, TSAT was an Network Centric relay satellite system integrated with wideband, protected and intelligence system. Five GEO satellites were connected by laser ISL to form constellation, interconnecting with AEHF, MOUS, WGS, APS and ORCA directly or indirectly, using laser or RF link communicating with other data satellites, airborne warning aircrafts, UAVs and ground stations.

The TSAT plan was eventually ran aground due to various reasons. So, in the view of construction and operation, each subsystem of the U.S. space-based system was not truly effective integrated, which greatly restricted the subsystem efficiency. However with the concept of TSAT proposing to wholly integrate scattered subsystems, it would provide a useful reference for the construction of our military NDSIN.

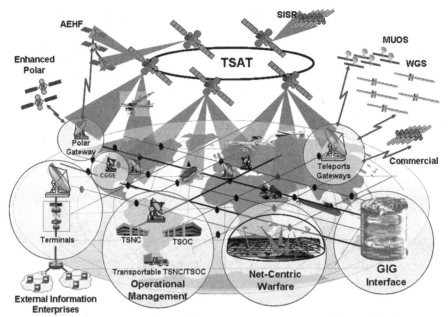

Transformational SATCOM System -- Operational View (OV-1)

Fig. 2. Schematic diagram of TSAT system [6]

4 The NDSIN Architecture

According to Chinese development strategy on space information network and missions of military operation other than war and missions of integrated joint operation in the future, the NDSIN, as the key point of national defense infrastructure construction, will be an information network that has regional expanding to global coverage, focused on system security, with the abilities of quick response, quick deployment and saturation information support, and could apply in government, industry sectors and modern military activities. Designing the total architecture of NDSIN, the factors should be considered as follow:

- The NDSIN will be an information network that has regional expanding to global coverage. When designing satellite constellation, the ability of global coverage should be considered.
- The NDSIN contains a variety of heterogeneous space platform, needing information gathering, fusion and forwarding to complete a variety of services, and must having the link transmission capability of large capacity space backbone. Therefore, inevitable requirement of on-board processing and switching should be considered.
- The NDSIN is a global coverage constellation, constricted by setting the abroad ground and TT&C stations. It requires the network running autonomously to solve the TT&C problem out of our country and to improve network survivability.

- The NDSIN are composed by satellite nodes with different orbits, types and properties. Dynamic and heterogeneous network structure must require standardization, normalization, informatization, flattening operation and management control mechanism.
- The national defense and military application requirements of NDSIN determine that the information security has become an important part of network construction, and it must have powerful security protection mechanism.

Therefore, under the analysis of space environment characteristics and the space-based platform ability, facing the different application requirements of national defense and military, a service oriented uninterrupted NDSIN architecture was designed. In Fig. 3, the schematic of NDSIN structure was given above. It was divided to three subsystems including space, operation and application segment. The space segment was composed of several satellites, which could be used to realize continuous or discontinuous coverage of the whole world or regions. Application segment referred to all kinds of user terminals, including handheld, vehicle (airborne, carrier) station, weapon platform terminal, space carrier terminal, portable stations and fixed stations, etc. The operation

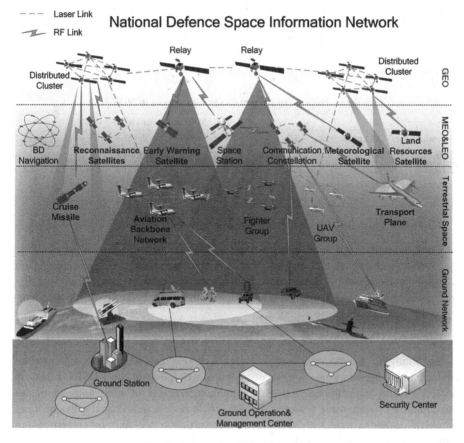

Fig. 3. Schematic of NDSIN structure

control segment was responsible for space and application segment, including network operation control subsystem, application management subsystem, safety protection subsystem, gateway, TT&C and their OSN (Operation Support Network).

5 Key Points to Be Broken

5.1 Topology Design and Optimization Method

According to the future space information network development strategy in our country, combining with the network architecture design idea, the NDSIN would be a regional expanding to global coverage information network system. Different from topology design of single type satellite subnet, the space segment of NDSIN related to satellite nodes with different orbit height, type and performance. All kinds of factors interact with each other and the constraint conditions were complex and changeable. The topology design problem would affect the overall performance and design cost of the network.

The basic characteristic of several typical satellite system of NDSIN was given in Table 2. It could be seen various types of services include most of the satellite orbit height, the number of nodes was huge, ISL requirements and coverage were different. All above resulted in the presence of traffic heterogeneity, space segment orbital layered, heterogeneous network topology and other characteristics, so design and optimization of space segment topology was one of the key technique [7].

Table 2. Basical parameters of satellite system in NDSIN

Service type	Orbit	ISL	Node num.	Coverage
Communication	LEO constellation + GEO	Yes	~ 100	Global + regional enhancement
Relay	GEO circle, distributed cluster	Yes	~ 3	Regional
Navigation	MEO constellation + GEO + IGSO	No	~ 30	Global + regional enhancement
Reconnaissance	LEO constellation, cluster	Yes	~ 100	Global

- Sub network topology design and optimization

Combining the mission of different types of satellite subnet, based on the coverage and operating characteristics of GEO, IGSO, HEO, MEO and LEO, researching on each subnet constellation design and topology optimization method, analyzing the impact of subnet topology on network efficiency and construction cost, lay the foundation for the NDSIN construction.

- Whole NDSIN topology design and optimization

According to the construction goal in coverage characteristics, comprehensive services, high-speed transmission, security and other aspects of the NDSIN, based on the information transmission and distribution requirement of each subsystem, the interconnection relationship between subnets should be researched firstly. Establishing the overall topology model of NDSIN, stage construction goal and the corresponding network topology should be put forward.

5.2   Link Transmission Technology

The NDSIN required supporting all kinds of information, requirements of real-time, reliability, average delay and delay jitter were variable. And, the transmission link of NDSIN had the characteristics of large spatial and temporal scales, large bandwidth delay product feedback, long cycle, complex electromagnetic environment, the relative motion between the nodes, cost to maintain stable link was huge. At the same time, under the effect of electromagnetic radiation and the load on the satellite, processing, storage, power consumption and other restrictions. Compared to the ground network system, the NDSIN link transmission technology constricted by more factors, it was more difficult to achieve. Therefore, research on the satellite link (satellite to ground, satellite to satellite) transmission technology was essential part to the construction of NDSIN. Considering the characteristics of the satellite link and transmission, the current application of common link and transmission protocol [8] using in NDSIN were shown in the following Table 3.

Table 3. Link and transport protocols analysis

Protocol	Topology	Sync	Competition/Collision avoid	Advantages	Drawbacks
CSMA/CA (RTS/CTS)	Distributed	No	Competition	Don't need sync.	High link load
TCP/UDP (AX.25)	Distributed	No	Competition	Don't need sync.	Poor operation in large noise and bandwidth constrained links
TDMA	Centralized	Yes	Collision avoid	High bandwidth utilization	Not suitable for large scale network
Half duplex CDMA	Centralized	Yes	Collision avoid	Low delay, high throughput	Near far effect and MAI have a great influence on the performance
LDMA	Centralized distributed	Yes	Competition/Collision avoid	Maximum channel utilization	The extendibility of network is poor and the time slot scheduling is difficult
Hybrid TDMA/FDMA	Centralized	Yes	Collision avoid	High bandwidth utilization, eliminate hidden and exposed node problems	Not suitable for dense and high load network
Hybrid TDMA/CDMA	Centralized	Yes	Collision avoid	Low delay, high throughput, Suitable for scalable, configurable small satellite missions	Strict synchronization requirements

- Research on communication mode based on space link transmission

With the information transmission needs of joint operations, services construction, tactical terminal, analyzing transmission link performance and cost of all kinds of communication standard in NDSIN, combined with the constellation distribution, coverage characteristics, atmospheric wave transparent window, electromagnetic compatibility and other factors, the NDSIN link transmission technology based on MIMO, cooperative communication, multi relay communication and other efficient communication mode would be researched.

- Research on data format of space link transmission

Based on data format proposed by CCSDS protocol[1], research on distributed data processing method, the NDSIN data format for a variety of information needs would be designed. For data relay, network control, information fusion, data compression, statistical analysis, decision support and other information type, considering the comprehensive efficiency, accuracy, receiving delay etc. of data transmission to formulate the corresponding data format, efficient sharing of data link and fast multiplexing/de-multiplexing were achieved. Based on the user characteristics of satellite networks, a data format optimization method based on the performance parameters such as resolution, coverage and positioning accuracy was proposed.

- Research on spatial multiple access technology

According to the orbit rule and the relationship between changes of visual, research on the multiple access modes as hybrid FDMA/TDMA for high speed link, designing algorithm of time and frequency multiplexing, the channel utilization would be improved, and the number of available satellite tracking antenna configuration would be reduced. The multi access mode of CDMA, and a spread spectrum strategy with low data rate and high reliability were studied in this paper. Combining multiple access characteristics of NDSIN, the CSMA access algorithm based on random access user would be researched, and the efficient collision detection and random backoff mechanism would be designed.

- Research on error control of space link

Research on inter or intra point to point error control strategy between the onboard and ground system of NDSIN, error correction, channel information feedback, automatic power control and error retransmission method under large spatial and temporal scales would be analyzed. Efficient HARQ scheme based on forward error correction and error retransmission would be researched. Analyzing the influence of spatial channel crosstalk and Doppler frequency shift on signal propagation in different ways of multiple access and modulation, an effective method to suppress channel crosstalk and compensate random fading would be proposed. Based on the characteristics of inter satellite communication links, such as satellite orbit perturbation, antenna tracking and pointing error and time window, the application of cross layer feedback method in error control would be studied.

---

[1] CCSDS standards. http://public.ccsds.org/publications/SIS.aspx.

5.3 Onboard Laser/RF Multi Granularity Hybrid Switching Method

Research on mechanism of multi granularity data stream aggregation and distribution in physical/link/network layer [10], Laser/RF hybrid switching and cross layer label normalization method would be studied. Study on network distributed traffic prediction model in lager time and space scales, and cross layer multi granularity scheduling algorithm of physical/link/network layer and hybrid dynamic allocation method under spatial environment, multi integration scheduling mechanism combined distributed scheduling and centralized scheduling would be build. Research on the implementation method of multi granularity switching, full optical switching with characteristics of signal format transparent and transient response, and grooming design method of polymerization technology and laser/RF heterogeneous link, a hybrid cross layer, multi system compatible and configurable witching method would be set up shown in Fig. 4.

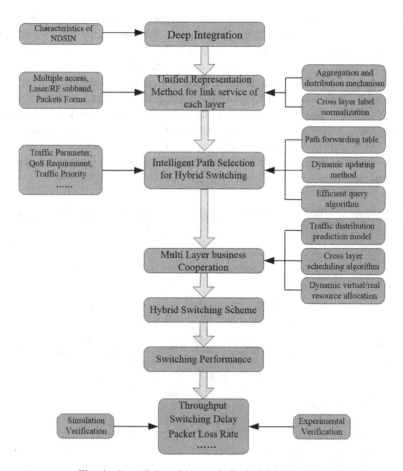

Fig. 4. Laser/RF multi granularity hybrid switching

5.4   Operation, Maintenance, Management and Control Technology

Analyzing system resource and application management of OMMCS (Operation, Maintenance, Management and Control System), research on function, composition and requirements for supporting network of OMMCS, the OMMCS structure should be studied from 6 aspects such as the network operation control, application management, safety protection, gateway station, TT&C station and OSN. The specific contents included two categories: one was the study on basic engineering facilities and reliable operation and maintenance of information network, responsible for communication control and management, to provide support for the user to carry out a variety of communication services; two was the user terminal access management and service acceptance, providing application support to operational communication departments.

5.5   Safety Protection Technology

Information security was a key point that any information system construction must be considered. The information security protection in the military field was a life-and-death matter determining the outcome of a war. In the NDSIN, a large number of information processing, analysis, transmission and receiving were strongly dependent on the space platform, space-based information system was facing great information warfare threat. Our country must vigorously develop the information protection technology of NDSIN, breaking the traditional information bottleneck of encryption, anti-interference, anti-interception, etc., must using the new concept, new technology to protect satellite information and its application. The NDSIN security protection diagram was shown in Fig. 5, which could provide confidentiality, integrity, availability, authenticity, non-repudiation and other security services for the NDSIN.

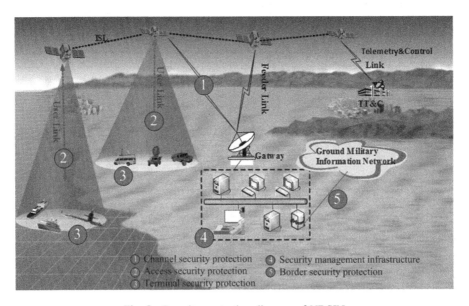

Fig. 5. Security protection diagram of NDSIN

6 Future Developing Trends and Frontier Technologies

Over the years, from "China's space" white paper[2,3,4] issued by P.R. China's State Council, space technology had become one of the most influential technologies in modern society, and the network construction in the field of aerospace technology had become the inevitable trend of the development of the world. The white paper also pointed out that the development long-term goal of China was "building space infrastructure with multiple functions and a variety of orbital tracks, composed of various satellite systems; building satellite ground application system and coordination, forming an integrated complete network system with continuous, long-term stability abilities". In order to ensure the rapid application and breakthrough of China's space technology, it is necessary to carry out the research on the related frontier technologies.

- Space optical communication technology [11]

Using light as an information carrier, compared with RF communication, the space optical communication had advantages of wide bandwidth, high transmission rate, good security, low power consumption, terminal miniaturization etc., would be the core of the future space communication. At present, the research focused on: optical domain processing and switching technology, laser communication ranging integration technology, laser/RF hybrid transmission technology and so on.

- SDN[5]

In 2008, the SDN (Software Define Network) was proposed by Stanford University Clean Slate Design [12], which made the transmission network and service network integrated together, and this direction began to enter a period of great development. The basic idea of SDN was: making the forwarded complex control logic information separate from the switch or router and other equipment in the network node (switching node), controlling information forwarding rules through software programming, finally achieving the purpose of free flow control. SDN technology was selected as one of the ten emerging technologies of "MIT" magazine in 2009. In order to simplify the complexity of information forwarding rules, the SDN defined a new network architecture based on the premise of not changing the information forwarding behavior of the original network.

- DTN[6]

DTN (Delay Tolerant Network) was mainly used to solve information transmission problems under the condition of the intermittent connection, great propagation delay, high bit error rate and extreme link asymmetry. Currently, the CCSDS had begun to deeply study BP (Bundle Protocol) protocol used in the field of aerospace application

---

[2] China's space 2000. http://www.gov.cn/gongbao/content/2001/content_61247.htm.

[3] China's space 2006. http://www.gov.cn/zwgk/2006-10/12/content_410824.htm.

[4] China's space 2011. http://www.gov.cn/gzdt/2011-12/29/content_2033030.htm.

[5] Stanford's Clean Slate Design for the Internet. http://clcanslate.stanford.edu.

[6] DTN Architecture. http://cwe.ccsds.org/sis/docs/SIS-DTN/dtnconcept paper.doc.

standard from 2012. The protocol could be used as the reference of our NDSIN network protocol design.

- QKD

The unconditional security of QKD (Quantum Key Distribution) scheme with its characteristics based on quantum mechanics and information theory framework, to assign the absolute security of key in communication between the two parties, could obtain a theoretically unbreakable secure encryption system with the "one-time" encryption algorithm. As the core of the solution, quantum secure communication with QKD could improve the existing IP network transmission link security level, and achieve unconditional security communication between users. Therefore, quantum key distribution technology had become the research hotspot in the field of information security.

- OAM [13, 14]

OAM (Orbital Angular Momentum) referred to the optical vortex beam contains a spiral phase factor, which has infinite eigenstates, the OAM of a single photon could carry infinite bit information in theory. The introduction of OAM into optical communication could greatly improve the optical communication capacity, which would become an important technology in Tbit optical communication.

7   Conclusion

The NDSIN was a complex network with large scale, heterogeneous topology and heterogeneous traffic. So, there were facing huge challenges in architecture, network model, network protocol, transmission mode, operation control, safety protection, performance analysis and other aspects. On the basis of previous research, the conception of NDSIN was put forward in this paper, to meet the urgent needs of the strategic support of China's future space-based information system, and then presented the network architecture design, analyzed its research direction and challenges. China initially had multi-track, multi-function and multi-type space-based systems, that could provide a necessary basis for the next step construction of NDSIN. However, the whole system construction was a huge and long project, must rely on the construction of relay backbone network, emphasized on laser/RF link hybrid transmission and processing mechanism, expanding research and application of the frontier technology gradually, completing the NDSIN construction step by step.

References

1. Rong-jun, S.: Some thoughts of Chinese integrated space-ground network system. Eng. Sci. **8**(10), 19–30 (2006). (in Chinese)
2. De-ren, L., Xin, S., Jian-ya, G., et al.: On construction of China's space information network. Geometrics Inf. Sci. Wuhan Univ. **40**(6), 711–715 (2015). (in Chinese)

3. Shi-quan, M.: An idea of China's space-based integrated information network. Spacecraft Eng. **22**(5), 1–14 (2013). (in Chinese)
4. Wei, H., Zhuang, L., Chao, D.: A study on the "One Belt and One Road" spatial information corridor. Ind. Econ. Rev. **5**, 125–133 (2015). (in Chinese)
5. Tarleton, R., Shively, S., Armstrong, B., Laurvick, C.: Transformational communications architecture for the department of defense, intelligence community and NASA. In: 24th AIAA International Communications Satellite Systems Conference, 11–14 June 2006, San Diego, California (2006)
6. Tie-feng, L., Ming-hua, Z.: The development of American military transformation satellite communication system. Satell. Netw. **5**, 62–63 (2011). (in Chinese)
7. Geng, Yu., Ding-rong, S., Shu-jian, L.: A spatial information network clustering algorithm. Acta Electron. Sinica **40**(3), 448–452 (2012). (in Chinese)
8. Pelton, J.N.: Satellite Communications. Springer, New York (2012)
9. Wang, C., Mao, K., Liu, J., Liu, J.: Identity-based dynamic authenticated group key agreement protocol for space information network. In: Lopez, J., Huang, X., Sandhu, R. (eds.) NSS 2013. LNCS, vol. 7873, pp. 535–548. Springer, Heidelberg (2013). doi:10.1007/978-3-642-38631-2_39
10. Chongxiu, Yu.: Optical Switch Technology. People's Posts & Telecommunication Press, Beijing (2008). (in Chinese)
11. Shanghong, Z., Yongjun, L., Jili, W.: Satellite Optical Networking. Science Press, Beijing (2010). (in Chinese)
12. Mckeown, N., Anderson, T., Balakrishnan, H., et al.: OpenFlow: enabling innovation in campus networks. ACM SIGCOMM Comput. Commun. Rev. **38**(2), 69–75 (2008)
13. Allen, L., Padgett, M., Babiker, M.: The orbital angular momentum of light. Prog. Opt. **39**, 291–372 (1999)
14. Xiao-cong, Y., Ping, J., Ting, L., et al.: Optical vortices and optical communication with orbital angular momentum. J. Shenzhen Univ. Sci. Eng. **31**(4), 331–346 (2014). (in Chinese)

Awareness of Space Information Network and Its Future Architecture

Chengwu Chang[1(✉)], Hongyang Liu[2], Liyu Cheng[1], and Jing Fan[2]

[1] Space Information Relay and Transmission Technology Research Center, Beijing, China
chang_cw@126.com
[2] Xichang Satellite Launch Center, Sichuan, China

Abstract. Entering to the 21st century, with the rapid development of information technology and the increasing expansion of mankind activity, The space information network (SIN) has become the key development field in the great powers around the world, the study of SIN shows an ascendant trend. Based on the analysis of the dynamic and developing trend of SIN at home and abroad, by studying the characteristics of SIN and its essential rules, and in the face of two kinds of developing tendency such as "Resilient and Decentralized Spatial Architecture" and "Integrated and Large-scale Spatial Architecture", this thesis proposes a new SIN architecture called "Space Information Cloud (SIC)", and has deeply studied the system structure of the public cloud in the space information cloud (SIC). In the end, the key technologies relevant to the space broadband backbone network were analyzed and discussed oriented to SIC. So as to provide reference about the development of SIN in China.

Keywords: Space Information Network (SIN) · Space Information Cloud (SIC) · Space broadband backbone network · Key technology · Future development

1 Introduction

Since entering the 21st century, with the rapid development of information technology and the expanding of human activity space, SIN has become the development of key areas which the powers are seeking to, SIN research shows an ascendant trend. The U.S. Military and National Aeronautics and Space Administration (NASA) successively put forward "Transformational Communication Architecture (TCA)" and "Space Communications and Navigation Network (SCaN)". In January 2013, Russia issued "Space Activities Planning for 2013-2020", which emphasized the development of space technology and equipment, to improve the space infrastructure. European countries launched "Allied Space-based Imaging System", "Galileo" system is expected to be put into use about 2020; In January 2013, Japan put forward "Universe Basic Plan", guiding the next 5~10 years' space development. All countries are in fighting for dominance in the field of SIN.

Research in the field of SIN in China began in "National High Technology Research and Development Outline" (namely "863 Plan"). In recent years, with the interests of the state to extend the global, SIN has become a common concern of aerospace experts

© Springer Nature Singapore Pte Ltd. 2017
Q. Yu (Ed.): SINC 2016, CCIS 688, pp. 80–92, 2017.
DOI: 10.1007/978-981-10-4403-8_8

and scholars around State and the Army. In addition to "973 Plan" and "863 Plan", in October 2013, the Ministry of Industry and Information Technology (MIIT) launched the "Space-Ground Integration Network" research program. In December 2013, the National Natural Science Foundation of China (NNSFC) launched the major research plan "Space Information Network (SIN)". What is SIN? What is the development rules of SIN? What kind of development way should be chosen for SIN? These are the studying hot issues.

Based on the analysis of domestic and foreign SIN research dynamic and its development trend, on the basis of studying the characteristics of SIN and the essential rule, and in view of these two trends of "Resilient and Decentralized Spatial Architecture" and "Integrated and Large-scale Spatial Architecture", put forward the new SIN architecture called "Space Information Cloud (SIC)", and studies the public cloud system structure of SIC in detail. Finally, analyzed the key technologies of space broadband backbone network oriented to SIC, in order to provide references on China's SIN development in the future.

2   Research Dynamic and Its Development Trend of SIN in China

In the 1990s, in order to track the development of space information systems beyond the pace of foreign countries, the "863 Plan" network research in China put forward the concept--"space-based integrated network". Since "the 10th Five--Year", the "863 Plan" and the "973 Plan" have also supported the SIN basic theory and mechanism research. In October and December of 2013, MIIT and NNSFC have also put forward major plans for the "Space-ground Integrated Network" and the "Space Information Network". The main research results are:

In the "863 Plan" network research domain, the main idea of "space-based integrated network" concept is: the future space-based integrated network is a type of network configuration that each node is equal in the network, and each node in the network plays multiple roles such as information obtainers, information transfer, and information processor. The satellite in the future is a set of multiple functions as an integrated satellite. From the network structure theory, the space-based integrated network is actually a peer-to-peer network. Under the guidance of the advanced concept of the space-based integrated network, the relevant units have carried out the basic theoretical research on various topics, such as: network topology research, multilayer network routing technology research, space network protocol research, and key components research for the space network. The concept of space-based integrated network led to the development of the basic theory of space network.

The main point of the "Space-ground Integrated Network" team of MIIT is: Space backbone network consists of GEO satellites, space access network is composed of space LEO satellites, through inter-satellite link between GEO backbone satellite and LEO access satellite, combines all the satellites into one whole network, which is connected with air-subnet and sea-subnet through inter-satellite links, and is connected with the Internet and access nodes on the ground through satellite-ground links, thus form a space-ground integrated network. As shown in Fig. 1.

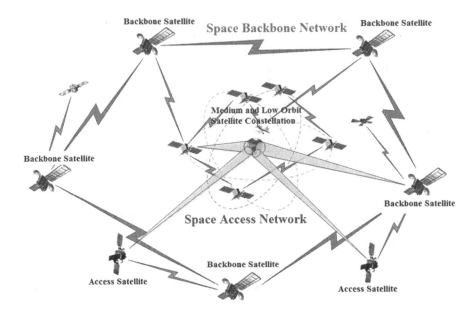

Fig. 1. Space-ground integrated network diagram

In terms of system architecture, the concept--"Space Information Network" proposed by NSFC is not fundamentally different from that of MIIT. The "backbone network + access network" architecture is proposed. But SIN is more focused on the basic theory researches, such as space-time datum, satellite communication network and navigation positioning enhancement research, resource allocation and operation managemen.

Throughout the space network architecture researches of the domestic research institutes, mainly in the following views: Firstly, the "backbone + access network + subnet /node" doctrine was led by Shen Rongjun academician, advocated all kinds of subnets /nodes through access networks, and then access the backbone network to form the space-ground integrated network; Secondly, the "backbone + autonomous subnet" theory proposed by Key Laboratory of Military Communications Satellite of PLA Science and Technology University, which is different from the "backbone + access network", directly access the backbone network; Thirdly, Professor Lu Jianhua from Tsinghua University proposed the "Rose Constellation" theory, for the geostationary orbits are scarce, using six satellites on MEO orbits to complete the global coverage; Fourthly, Beijing Space Information Relay and Transmission Technology Research Center, relying on Key Laboratory of Areospace Broadband Network, proposed many research points such as the "dual-ring structure of laser and microwave" and "backbone + aggregation + access three-layer model" on the space backbone network, which is deepening the research of the space backbone network.

While the SIN theory continues to advance, the key components' research and system simulation experiments of SIN also made progress. Space Center of CAS (Chinese Academy of Sciences), National University of Defense Technology, Beijing University of Aeronautics and Astronautics and other units have developed the space router and space gateway; Dalian University and other units studied space network protocol, and established a simulation system; Based on "Space Integrated Information System Technology" National Defense Key Laboratory, Software Institute of CAS has carried out the theoretical research on the software-defined spac network, and has completed mathematical modeling and system simulation for all kinds of orbiting spacecraft. The Fifth Academy of China Aerospace Science and Technology Group has successfully developed the aerospace product prototypes of Ka multi-beam antenna, space multi-protocol hybrid router and 100 Mbit/s satellite modulation- demodulation module, etc. The Eighth Academy of China Aerospace Science and Technology Group has accumulated rich experience in the research of new-generation agile exploration satellites, information processing and storage technology on orbit, which has been successfully applied on satellites. Xi'an Institute of Optics and Fine Mechanics, Harbin Institute of Technology, Dalian University of Technology, Air Force Engineering University and other units are studying space laser communication, optical switching network, routing strategy research work, and have achieved many results.

3   Framework of SIN Architecture and Its Future Development

3.1   Further Understanding on the Basic Characteristics of Network Structure

According to the basic theory of network, there are two typical types of network structure: One is the peer-to-peer network, and the other is the backbone-access network. As shown in Figs. 2 and 3.

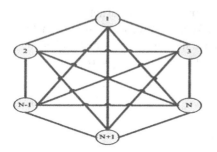

Fig. 2. Peer -peer network diagram   **Fig. 3.** Backbone-access network diagram

A peer-to-peer network is a distributed network, each node and other nodes are connected, can also be called non-central network. A backbone-access network is a centralized network, other nodes or subnets access the backbone network in order to interconnect and interwork. In addition to a few autonomous networks, terrestrial networks is backbone-access networks, such as China's network, which consists of "eight vertical and eight horizontal" optical fibers to form a backbone network, other users and subnets access it by core nodes; Internet is a typical backbone-access network, among the continents through submarine cables' connection, to form a global Internet; GIG is also a backbone-access network, through the GIG-BE (fiber expansion program) and satellite broadband networks to achieve the interoperability of the theater combat unit.

In the development of space network, there are two types of networks, namely, peer-to-peer and backbone access networks. Iridium is a typical peer-to-peer network, each node is connected with its adjacent nodes, there is no central node. The US Military Transformational Satellite Communication System (TSAT) is a backbone- access network, to form a high-rate broadband ring network connected by 40 Gbit/s high-rate inter-satellite links, which can provide dozens of high-rate laser links and more than 8000 low-rate RF links for access services of other nodes and subnets. As shown in Figs. 4 and 5.

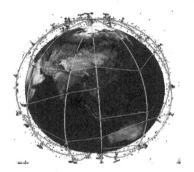

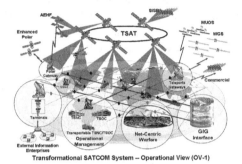

Fig. 4. Iridium satellite system diagram

Fig. 5. TSAT system network diagram

From the general network type and the development of terrestrial networks and space networks, backbone-access network is the main form of network, peer-to-peer network is generally applied in sub-network or professional network. The future SIN belongs to hybrid network structure as the main body of backbone-access network.

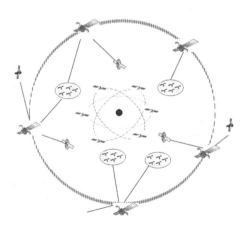

Fig. 6. Space information network structure diagram in the future

The future SIN will be a heterogeneous network in which peer- peer and backbone-access networks co-exist. For a autonomous system with independent functions, such as LEO satellite communication system, early warning satellite system, navigation and positioning system, etc., the mode of peer-to-peer network can be maintained. For large data remote domain transmission and SIN to support the vast majority of application mode, will remain the backbone-access network model. As shown in Fig. 6.

Figure 6 shows, the intermediate layer satellite indicates a peer-to-peer satellite network, but the backbone-access network indicated by the outermost layer, the subnet of the backbone network access may be a autonomous peer-to-peer network or a second core-subnet. Therefore, the future SIN is a complex heterogeneous network consisting of peer-to-peer network and backbone-access network.

3.2   Research of SIN Must Grasp the Space-Time Law of Changing and Unchanged

Network as a complex system, need to support multi-user information sharing and exchange, we must balance the reliability and flexibility.

From terrestrial network architecture, terrestrial network consists of core networks and access networks, a core network is generally SDH optical network, an access network is generally an integrated access network, as shown in Fig. 7.

From terrestrial network protocol system, core networks are circuit-switched networks based on wavelength, using SDH link layer with a unified data unit format, access networks are end-to-end communication connections, adopting IP exchange model.

Terrestrial networks are the composition of core networks based on circuit-switched and access networks based on packet switching, IP packet switching is an unreliable mode of data transmission, it is precisely high reliable because of core network circuit switching, to ensure end-to-end transmission quality.

From the space-time law of SIN, SIN is a large space-time, long delay, interrupted, asymmetric, time-varying network, which is different from terrestrial networks. Otherwise, the main nodes of SIN are spacecrafts, the movement of spacecrafts is regular, especially GEO satellites, relative to the earth static, this is the constant part. The future development of SIN, can learn from the inherent law of the development of terrestrial networks, based on GEO satellites to build a core network, using laser-based wavelength switching to construct a high-rate, stable and reliable space backbone network, data aggregation, fusion processing, space-time benchmarks and other subnets or high-rate nodes directly connect with the core network or through the access network interconnection, various professional subnet on demand can be used packet switching, virtual circuit switching, circuit switching and various types to solve the conflict among the differences of services, the flexibility of application and distribution of the reliability.

SIN based on GEO satellites is a typical backbone-access network because of its extremely high broadband and layered access characteristics. Its overall system architecture is shown in Fig. 8.

In this network architecture, the high-rate communication transmission node and the sub-network main node directly connect with the laser broadband network through laser

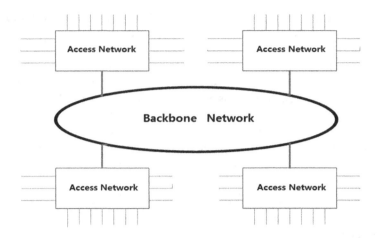

Fig. 7. Terrestrial network structure diagram

links, and other adjacent sub-networks, air-based subnets and sea-based subnets are indirectly connected to the laser broadband network via space access networks, to achieve remote transmission and information sharing.

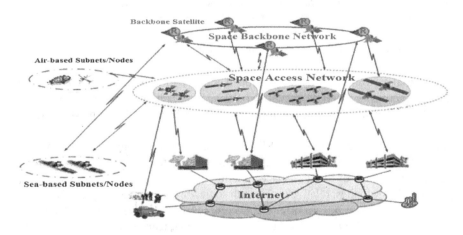

Fig. 8. Structure of space information networks based on GEO satellites

3.3   Foundation and Structure of SIN for Space Information Cloud (SIC)

3.3.1   Space System Architecture Development Trends and Our Views

On the one hand, US Center for Strategic and Budgetary Assessment released the report "The Future of US Military Communications Satellites" on July 24, 2013, proposing the "Separation, Decentralization or Expansion of the Spatial Structure of the System". In August 2013, US Air Force released the paper "The Resilient and Decentralized Spatial Architecture", proposing a decomposable, deformable and reconfigurable architecture to distribute spatial tasks and functions into different systems across multiple orbits and platforms. US Military's "See Me", "F6" Systems are typical distributed systems, showing a small, decentralized development trend. On the other hand, US Military's "MUOS", "AEHF", "WGS" are large capacity satellites, such as WGS equipped carried by GBS (Global Broadcasting System), US Satellite Communications Company's communication satellites, using KaMA multi-beam technology, the capacity of a single satellite is 140 GB. Japan "Quasi-zenith" satellite system will be integrated navigation and communications, showing a large-scale, comprehensive development trend.

Through research, we believe that: these two trends are related to different needs, there is no substitute for the relationship between "Resilient and Decentralized Spatial Architecture" and "Integrated and Large-scale Spatial Architecture", which will be long-term coexistence and coordinated development. The resilient and decentralized spatial architecture emphasizes the survivability of space systems, hoping to have the minimum communication and situational awareness capability to protect the system against physical attacks in an environment of space attack and defense. The integrated and large-scale spatial architecture emphasizes space systems function's completeness, with the strong information support and interoperability capability for the coming information operations. We believe that these two trends will be unified in the development of SIN.

3.3.2   SIN Architecture Oriented to Spatial Information Cloud

Cloud Computing as a unique IT service model, CSA (Cloud Security Alliance) released the book "Security Guidance For Critical Areas of Focus in Cloud Computing V3.0", which described precisely the nature of cloud computing--"The nature of cloud computing is a service delivery model, which can be shared resource pools on demand at anytime and anywhere. Resource pools include computing resources, network resources, storage resources, which can be dynamically allocated and adjusted, and flexibly divided in different users." In order to more easily match this definition, NIST (National Institute of Standards and Technology) put forward the definition of cloud computing standards –"NIST Working Definition of Cloud Computing /NIST 800-145". This document refers to a standard cloud computing needs to have five basic elements, namely: network distribution services, self-service, measurable services, flexible scheduling of resources, and resource pooling.

The architecture of SIN oriented to Space Information Cloud (SIC), based on the cloud computing architecture and the application and development of SIN, the following tentative ideas are put forward, as shown in Fig. 9.

(1) *Basic Elements of SIC Services.*
    (1) Self-service: Members of SIC, can receive the space-time datum and customized information products services on demand.
    (2) Network Distribution: Based on network distribution and SIC, to provide computing or storage services for the members everywhere.
    (3) Resource Pooling: Resource pooling refers to shielding the difference of all types of spacecrafts, considers each spacecraft as a data, storage, computing,

communication carrier, and considers resource virtualization as the SIN computing basis.

(4) Flexible Scheduling: On the basis of resource pooling, to achieve the flexible configuration of resource allocation and scheduling.

(5) Resource Balance: In order to improve the efficiency of SIN, on the basis of resource pooling, resource scheduling and resource aware, to implement resource balancing among computing, communication and storage.

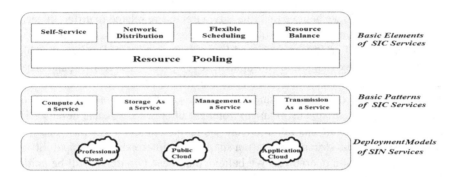

Fig. 9. Space information network architecture oriented to space information cloud

(2) *Basic Pattern of SIC Service.*

(1) Computing-as-a-service: In SIC computing, including information fusion processing, spacecraft orbit calculation, spacecraft mutual perception and configuration maintenance, SIC provides standardized computing services for all kinds of members.

(2) Storage-as-a-service: In SIN, a large amount of distributed storage resources are needed because of the interruption characteristics of spatial links and the original data and information products storage.

(3) Management-as-a-service: Due to the dynamic characteristics of spacecrafts, SIN needs to provide the management services of dynamic access, network self-organization and resource application for each member.

(4) Transmission-as-a-service: The efficient operation of SIN is based on spatial links, link establishment, link maintenance and spatial data transmission are the basic service capabilities of SIN.

(3) *Deployment Model SIN Service*

(1) Public Cloud: refer to the public resources in SIN with communication, storage, computing functions, generally deployed in large spacecrafts and the space broadband backbone network.

(2) Professional Cloud: refer to the professional resources with space-time datum, detection awareness, multi-access functions, generally composed of professional subnets.

(3) Application Cloud: mainly refer to the space information cloud service object, which are members to accept information service in SIC.

3.3.3   Common Foundation of SIN Oriented to Space Information Clouds

The architecture of space information cloud computing is related to computing, storage, management and so on. It involves the interrelationship of professional cloud, public cloud and application cloud, which needs a stable and high-rate broadband network to integrate these compositions and services. Space backbone networks, based on GEO satellites, as the public infrastructure of SIN, will play interconnection, storage, and computing functions.

(1) *Public Cloud System Components of SIC Based on GEO Satellites.*

The space information backbone network, based on GEO satellite, is a public cloud of SIN, and also is a space-ground integrated backbone network, which is combined with data network and management network, and is a complex network system that other subnets and nodes can directly connect through links or access-networks, as shown in Fig. 10.

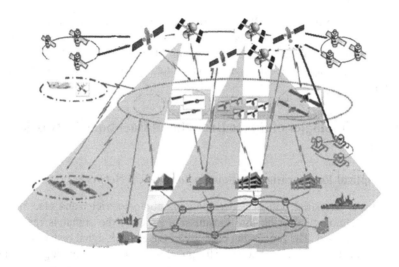

Fig. 10. Public cloud composition diagram of spatial information cloud

(1) Space Data Network: the space backbone network is composed of GEO communication satellites, through inter-satellite laser communication links to complete remote distribution and information exchange for other professional applications subnets and nodes.
(2) Space Management Network: the ubiquitous space network is composed of SMA digital multi-beams carried by GEO satellites, to accompolish the seamless connection within MEO, LEO spacecrafts and near-space, air-based, ground, sea-based valuable objectives, to complete the management of all objects.

(2) *Public Cloud Nodes' Structure of SIC Based on GEO Satellites*

In order to improve the system capacity and reliability of the space information backbone network, it is possible to design the node configuration in the following manner.

(1) Co-location

As a space information backbone network, co-location can achieve three purposes: similar satellite emergency backup, similar satellite capability overlay and different satellite complementary functions.

(2) Distribution

In order to solve the problem of insufficient capacity of the single satellite, the distributed satellite solution can be considered, through the space standard interface, to form a distributed satellite with connected, processed and exchanged functions.

(3) Integration

In order to improve the satellite capability of single GEO orbit, it is possible to construct integrated functional satellites by software functional reconstruction and multi-load integration to meet the requirements of high-rate data transmission and distribution, data collection, communication protocol conversion and high rate data processing.

4 Key Technologies of Space Broadband Backbone Network Based on SIC

4.1 All-Optical Regeneration Technology for Optical Network

At present, the terrestrial optical network adopts photoelectricity and electro-optical conversion technology. During the transmission process, the signal is regenerated, shaped and amplified. The processing capacity of the electric signal restricts the transmission rate of the backbone network. All-optical regeneration technology for optical network, especially for the space network, is very necessary because of the space-based platform processing capacity constraints and the level of anti-radiation device.

4.2 High-Order Modulation and Demodulation Technology of Space Laser Communication

At present, in terrestrial optical networks, laser high-order modulation and demodulation technology is very common. However, in the space optical network, due to the influence of phase, high-order modulation and demodulation technology, has not been fully applied. In space optical communication, high-order modulation and demodulation to enhance single-wavelength link capacity, in particular, is necessary to study.

4.3 High Sensitivity Detection of High-Rate Weak Optical Signals Under WDM

WDM technology has been applied successfully in terrestrial optical networks. However, due to the influence of phase jitter and so on, in space optical communication, if adopting wavelength division multiplexing (WDM) technology, it is necessary to solve the high sensitivity of high-rate weak optical signal detection technology.

4.4 Space Laser Communication Efficient Protocol Design Technology

In terrestrial optical communication, mostly adopt SDH protocol to carry IP, ATM and other protocols. In space communication, there are many protocols such as CCSDS, IP, DTN and ATM. At optical communication level, data flow or CCSDS protocol is mainly used, SDH protocol for the feasibility of space optical communication, is also need to study.

4.5 Space Heterogeneous Network Interconnection and Aggregation Technology

Due to different link types and different protocols, SIN expresses the characteristics of heterogeneous networks, but the nature of the network is interoperability and information sharing. It is necessary to solve the interconnection of heterogeneous networks and the aggregation of multichannel low-rate datas. At present, SIN research is more and more, but for heterogeneous network interconnection, most of solutions are the use of gateway for protocol conversion. This approach can be used for single-link, but for space network it isn't high efficient! Fundamentally, it is needed to innovate the method of space heterogeneous network interconnection and multichannel aggregation.

4.6 Space Optical Network Synchronization Technology

In the space optical network, as with the terrestrial network, there are also time-frequency synchronization problems. The terrestrial optical network uses a time synchronization network to complete time-frequency synchronization. For the space network, they are worthy of deepening research to adopt the independent time synchronization network, or the channel associated signaling, or the Beidou signal.

4.7 Space Laser Communication Digital and Analog Hybrid Transmission Technology

From the future development trend of space technology, space information exchange will be mainly composed of circuit switching and packet switching. In the backbone network, there will coexist in high-rate data analog modulation and multiplexed digital modulation. The realization of hybrid transmission and exchange including digital signal and analog signal, is the topic of in-depth study.

5   Conclusion

SIN is a comprehensive discipline involved in many disciplines such as orbital dynamics, spacecraft design, constellation theory, satellite communication, communication network, data processing, remote sensing, control engineering, and complex system theory. Many key technologies need to be studied. But from the network point of view, the essence of the network is to link all the members of the system, so as to achieve information sharing and interoperability among members. The most basic issues are two: One is heterogeneous network interconnection problem, mainly caused by different service objects, different application scenarios or different network protocols, involving network switching technology and protocol system; The other is the network management issue, in particular, to realize real-time management of hundreds of spacecrafts, involving access technology, management mode and spatial target mobility management and other technical issues. The research of these two basic technologies is related to the network demand, the development of technology and the mode of operation and management. Therefore, the research of SIN has a systematic, challenging and multi-directional characteristics. In this paper, the SIN architecture and its key technologies oriented to SIC are only some thoughts and discoveries in the process of SIN research, we hope to provide some reference for people to study SIN.

References

1. Qing, P., Xinjie, H., Xiaoqing, Z.: Network Centric Warfare Equipment System, pp. 47–48. National Defense Industry Press, Beijing (2010). (in Chinese)
2. Shiquan, M.: Satellite Communication System Engineer Design and Application. Publishing House of Electronics Industry, pp. 517–518 (2015). (in Chinese)
3. Haibin, W.: A Study on Russian Revolution in Military Affairs of Putin Period. China Social Sciences Publishing House, pp. 155–156 (2010). (in Chinese)
4. Lina, W., Bing, W.: Satellite communication system. National Defense Industry Press, Beijing, pp. 323–324 (2014). (in Chinese)
5. Rongjun, Shen: Some Thoughts of Chinaese integrated space-ground network system. Eng. Sci. **8**(10), 25–26 (2006). (in Chinese)
6. Chengwu, C.: Research on aerospace information network based on relay satellite node. Space Information Relay and Transmission Technology Research Center, Beijing, Technical report: 2015AA7011071 (2015)
7. Li, Z., Jielin, F., et al.: Modern communication network and its key technology. p. 9, National Defense Industry Press, Beijing (2011). (in Chinese)

Theory and Method of High Speed Transmission in Space Dynamic Network

A Congestion-Based Random-Encountering Routing (CTR) Scheme

Chao Zhao[1]([✉]), Gang Xie[2], Jingchun Gao[2], and Kaiming Liu[2]

[1] Beijing University of Posts and Telecommunications, Beijing Antuosi Technology Co. Ltd.,
Beijing, China
[2] Beijing University of Posts and Telecommunications, Beijing, China
{xiegang,gjc,kmliu}@bupt.edu.cn

Abstract. This paper proposes a routing scheme for low earth orbit (LEO) satellite networks, named congestion-based random-encountering routing (CTR) scheme. Firstly, forward agents migrate autonomously to explore paths from source to destination satellite, while back agents follow the path in the opposite direction. Meanwhile, data probability table of every satellite will be changed according to the condition of satellites which agents gather and estimate, eg. congestion, propagation and queuing delay between inter-satellite link (ISL). Finally, simulation is performed on a Courier-like constellation. Moreover, results show that the proposed scheme guarantees better quality of service (QoS), especially shorter end-to-end delay and lower packet loss ratio, for load balancing.

Keywords: Routing scheme · LEO satellite networks · Probability table · Quality of service · Load balancing

1 Introduction

Satellite networks will be an important role of the next generation Internet for the capability of global coverage, inherent multicast and consistent service. Compared with the geostationary orbit (GEO) satellites and medium earth orbit (MEO) satellites, LEO satellites attract increasing attention for shorter propagation delay, less transmission loss and high bandwidth capability. With the development of multimedia in the Internet, traffic QoS in LEO satellite networks should be always guaranteed. Since population distribution and economic development are highly unbalanced on the surface of the earth, traffic requirements are unfair in LEO satellite networks. Most hot spots, such as Asia, Western Europe and Northern America, are located in the northern hemisphere, above which satellites are usually congested while others are idle. Given large number of packets dropping at congestion areas, a flexible and efficient routing scheme should be designed for research.

So far, many routing scheme have been developed for LEO satellite networks. A reverse detection based QoS routing algorithm [1] was proposed on the dynamic topology of LEO satellite networks to choose the least-delay path to transmit traffic. But single path fails to handle an abrupt congestion event, which will results in high packet

© Springer Nature Singapore Pte Ltd. 2017
Q. Yu (Ed.): SINC 2016, CCIS 688, pp. 95–103, 2017.
DOI: 10.1007/978-981-10-4403-8_9

drop rate. In [2, 3], the explicit load balancing (ELB) scheme was developed to address congestion in satellite networks. In the scheme, a satellite is notified by its neighbor satellite with high traffic load to reduce data transmission rate, and the reduction portion of traffic chooses an alternate path to the destination satellite. ELB with encouraging results of better traffic distribution, higher throughput, and lower packet drops is tailored for double-layered satellite networks [4] and receives desirable results. Even, ELB scheme was adopted in [5], which guarantees traffic balancing. Based on the idea of ELB scheme, a traffic-light-based intelligent routing (TLR) strategy was developed in [6]. TLR strategy considers two hops including current and next hop to choose a better router. Compared to ELB and TRL routing scheme, several algorithms try to positively explore more paths rather than passively avoid congestion. A distributed routing algorithm with traffic prediction (TPDRA) [7] used traffic prediction and routing decision two parts, achieving better load balancing. In [8–11], agents are used to explore paths and update routing table. When it comes to updating routing table, delay and geographical factor are used to decide next hop, which gets better network QoS. In [12], physical layer is used to offer information to network layer when making routing decision and it has improved the robustness of satellite networks.

In this paper, a congestion-based random-encountering routing (CTR) scheme is proposed. The idea of ELB scheme is implemented. An ISL sate can be divided into 3 stages and 2 thresholds are calculated according to the condition of satellites. Agents explore a feasible path and update the probability table. A cost function is also used in updating the probability table.

2   Framework for the Model

As is shown in Fig. 1, LEO satellite network consists of N planes, M satellites in each plane. An ISL between two satellites in the same plane is named as intra-plane ISL, while in different planes is named as inter-plane ISL. In addition, the links between ground terminals and satellites are named as UDLs. The propagation delay on intra-plane ISLs is always fixed. But the propagation delay on inter-plane ISLs is variable as the distance between two satellites in different planes always changes with latitude. The propagation delay on intra-plane ISLs and inter-plane ISLs is respectively about 13.47 ms and 11.58 * cos (lat) ms [11], where lat is satellite latitude. The inter-plane ISLs are closed when satellites are above the Polar Region. When two satellites in adjacent planes move in opposite direction, there will be cross-seams between them. For simplicity, we don't consider cross-seams.

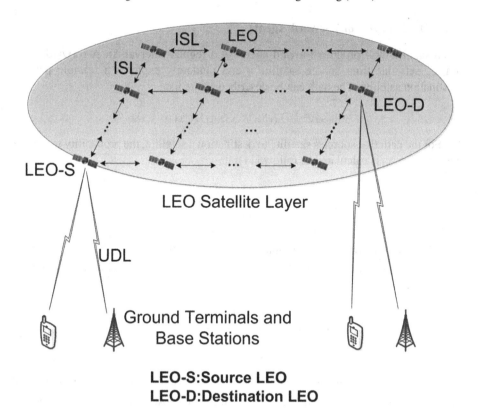

Fig. 1. The LEO satellite network

LEO satellite network topology can be described as a graph $G = (V, E)$, where V and E respectively represent the set of all satellite nodes and all existing ISLs. $|V|$ and $|E|$ are respectively the number of satellite nodes and ISLs, and $|V| = N \times M$. And $N(i)$ is s set of all neighbor satellites of satellite i.

3   The Proposed Method

We present a congestion-based random-encountering routing (CTR) scheme. The scheme aims at avoiding congestion adaptively, which not only achieve loading balancing but also guarantees QoS of the entire satellite network. Agents, including forward agents and back agents, and are used in the scheme. Forward agents are used for paths exploration and information collection, and back agents with information of forward update satellite probability table.

3.1   The Behavior of Forward Agents

Each satellite will produce forward agents at a regular interval Δt. A forward agent $FA_{s->d}$ sets out from source satellite s and randomly explores a feasible path to destination satellite. The path can be described as the list:

$$Path_{s->d} = [s, v_1, v_2, \cdots, v_{k-1}, v_k, v_{k+1}, \cdots, v_n, d]$$

For the path that source s satellite to destination d satellite, the probability to choose next satellite v_1 is calculated as follow [11]:

$$(P_{s,d}^{v_1})_{agent} = \frac{1/hop_{v_1,d}}{\sum\limits_{j \in N(s)} 1/hop_{j,d}} \tag{1}$$

Where $hop_{v_1,d}$ donates the minimum hops from satellite v_1 to satellite d.

When it comes to an intermediate satellite v_k, the probability to choose next satellite v_{k+1} is calculated as follow:

$$(P_{v_k,d}^{v_{k+1}})_{agent} = \frac{(P_{v_k,d}^{v_{k+1}})_{data}}{\sum\limits_{j \in N'(v_k)} (P_{v_k,d}^{j})_{data}} \tag{2}$$

Where $(P_{v_k,d}^{v_{k+1}})_{data}$ donates the probability that data from satellite v_k to d chooses next hop v_{k+1}. $N'(v_k)$ donates next hop satellites that can be chosen. In case of endless hoops, we abandon the agent that arrives at a satellite if its neighbor satellites have been all visited by itself.

3.2   The Behavior of Back Agents

When $FA_{s->d}$ arrives at satellite d, it is abandoned and a back agent $BA_{s->d}$ is produced. $BA_{s->d}$ follows $Path_{s->d}$ in the opposite direction. Once $BA_{s->d}$ gets intermediate satellite v_k, the probability table is updated. The probability table storages the probability that a satellite choose next hop for a given destination d.

The probability of v_k choosing v_{k+1} which belongs to $Path_{s->d}$ will be updated as Eq. (3).

$$(P_{v_k,d}^{v_{k+1}})_{data} = \begin{cases} (P_{v_k,d}^{v_{k+1}})_{data} + (1 - (P_{v_k,d}^{v_{k+1}})_{data}) \cdot \gamma & \delta_{v_k,v_{k+1}} \leq \alpha \\ (P_{v_k,d}^{v_{k+1}})_{data} \cdot \exp(-\frac{1}{\beta - \delta_{v_k,v_{k+1}}} \cdot \eta) & \alpha < \delta_{v_k,v_{k+1}} \leq \beta \\ 0 & \delta_{v_k,v_{k+1}} > \beta \end{cases} \tag{3}$$

For $v_{k+1}' \notin Path_{s->d} \cap k \in N(v_k)$, if $\delta_{v_k,v_{k+1}}' > \beta$, the probability will be updated as Eq. (4).

$$
(P_{v_k,d}^{v'_{k+1}})_{data} = \begin{cases} (P_{v_k,d}^{v'_{k+1}})_{data} + (1 - (P_{v_k,d}^{v'_{k+1}})_{data}) \cdot \gamma & M \text{ in } cost_{v'_{k+1}} \\ (P_{v_k,d}^{v'_{k+1}})_{data} & \delta_{v_k,v'_{k+1}} \leq \alpha \\ (P_{v_k,d}^{v'_{k+1}})_{data} \cdot \exp(-\dfrac{1}{\beta - \delta_{v_k,v'_{k+1}}} \cdot \eta) & \alpha < \delta_{v_k,v'_{k+1}} \leq \beta \\ 0 & \delta_{v_k,v'_{k+1}} > \beta \end{cases} \tag{4}
$$

If $\delta_{v_k,v_{k+1}}' \leq \beta$, then the probability will be updated as Eq. (5).

$$
(P_{v_k,d}^{v'_{k+1}})_{data} = \begin{cases} (P_{v_k,d}^{v'_{k+1}})_{data} & \delta_{v_k,v'_{k+1}} \leq \alpha \\ (P_{v_k,d}^{v'_{k+1}})_{data} \cdot \exp(-\dfrac{1}{\beta - \delta_{v_k,v'_{k+1}}} \cdot \eta) & \alpha < \delta_{v_k,v'_{k+1}} \leq \beta \\ 0 & \delta_{v_k,v'_{k+1}} > \beta \end{cases} \tag{5}
$$

Where γ, η are real numbers, and $0 < \gamma < 1$. $\delta_{v_k,v_{k+1}}$ donates ISL congestion between satellite v_k and v_{k+1}. α, β donate two congestion thresholds, which are discussed next. $cost_{v'_{k+1}}$ is the cost between v_k and v_{k+1}.

And for $i = v_{k+1} || i = v'_{k+1}$, $(P_{v_k,d}^{i})_{data}$ is initialized as follow [11]:

$$
(P_{v_k,d}^{i})_{agent} = \frac{1/hop_{i,d}}{\sum\limits_{j \in N(v_k)} 1/hop_{j,d}} \tag{6}
$$

After updating the probability table, the probability should be normalized as follow:

$$
(P_{v_k,d}^{i})_{data} = \frac{(P_{v_k,d}^{i})_{data}}{\sum\limits_{j \in N'(v_k)} (P_{v_k,d}^{j})_{data}} \tag{7}
$$

In our work, cost is the sum of propagation delay and queuing delay between two adjacent satellites. Namely,

$$
cost = PD + QD. \tag{8}
$$

Where PD and QD are respectively propagation delay and queuing delay. The propagation delay can be calculated due to deterministic and periodic network topology. And the queuing delay can be calculated as Eq. (9) [9].

$$
QD = Num_{queue} \cdot \frac{P_{avg}}{C_{ISL}} \tag{9}
$$

Where Num_{queue}, P_{avg} and C_{ISL} are respectively packet queuing number, average packet size and ISL capacity.

3.3   Two Congestion Thresholds

Two congestion thresholds are calculated as ELB [2, 3] scheme. Hence,

$$\beta = 1 - Min(1, \frac{(\delta + \text{cost})(I - O)}{(Q_l - q) \cdot P_{avg}}). \qquad (10)$$

Where δ is satellite monitoring interval. I and O are respectively total input and output traffic rates of a satellite. Q_l and q are respectively the total length of a satellite and the occupancy of its queue. And

$$\alpha = \frac{\beta}{2}. \qquad (11)$$

4   Experimental Evaluation

To compare with LBRP-MA1 [9], we choose Courier as satellite constellation, with 8 planes and each plane with 9 satellites. $\gamma = 1, \eta = 1$ are adopted in study. Other parameters of LEO satellite network are listed as Table 1.

Table 1. Simulation parameters

Latitude threshold	80°N and 80°S
ISL queue type	FIFO
ISL queue length	100 Packets
ISL traffic rate	10 Mb/s
Packet size	500 B

To evaluate end-to-end delay and packet loss ratio, two pairs of terrestrial source-destination are chosen in study. The first pair source and destination are respectively located at (38°E, 18°N) in Oceania and (142°E, 38°S) in Africa. The second pair source and destination are respectively located at (71°W, 33°N) in North America and (122°E, 38°N) in Asia. The latter areas are at higher traffic load due to more population and faster economic growth.

Figures 2 and 3 show the average end-to-end delay and the average packet loss ratio versus load of the first pair. The end-to-end delay of the CTR scheme averagely reduces about 30% than LBRP-MA1. Because data traffic hardly takes a detour when the background flow is light in the CTR scheme. And the packet loss ratio is also lower in the CTR scheme, since the scheme mainly aims at congestion.

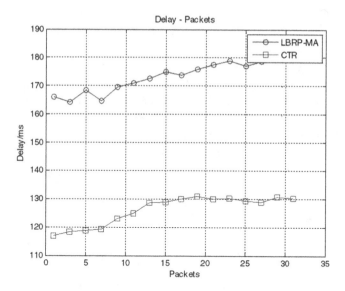

Fig. 2. Average end-to-end delay versus load of the first pair.

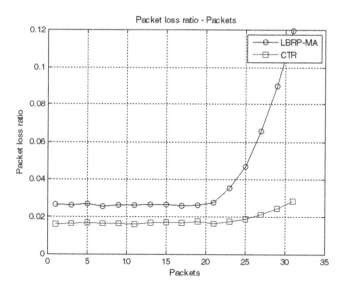

Fig. 3. Average packet loss ratio versus load of the first pair.

Figures 4 and 5 show the average end-to-end delay and the average packet loss ratio versus load of the second pair. The end-to-end delay of the CTR scheme is lower at low load; conversely it's higher than LBRP-MA1 at high load. Because data traffic will take more detours when the background flow is heavy and the load is heavy to avoid congestion in the CTR scheme. However, the packet loss ratio is always lower in the CTR scheme, especially when load is high. The results show the ability of load balancing in the CTR scheme.

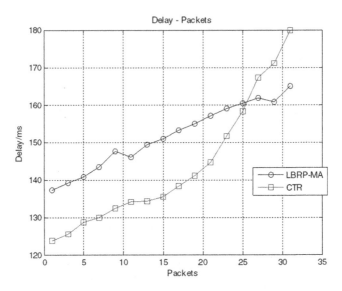

Fig. 4. Average end-to-end delay versus load of the second pair

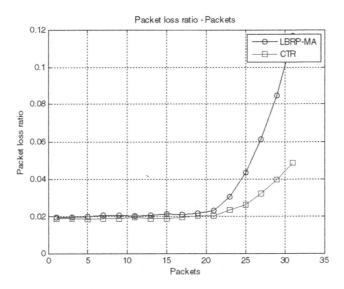

Fig. 5. Average packet loss ratio versus load of the second pair.

5   Conclusion and Future Work

In this paper, we present a congestion-based random-encountering routing (CTR) scheme in LEO satellite networks. According to ISL congestion ratio of three states with two thresholds, the scheme selects next hop to positively alleviate congestion, which has achieved load balancing in the networks. The simulation results show that the scheme gets low end-to-end delay and packet loss ratio, guaranteeing QoS of the networks.

In the future work, cross-seams will cause much cost and it's a difficult problem need to take in consideration. In addition, to guarantee reliability of the network, satellite failing is worth discussing in study.

Acknowledgment. This work was supported in part by Electronic Information Industry Development Fund and National Natural Science Foundation of China [grant number 61531007].

References

1. Yan, J., Zhang, Y., Cao, Z.: Reverse detection based QoS routing algorithm for LEO satellite constellation networks. Tsinghua Sci. Technol. **16**(4), 358–363 (2011)
2. Taleb, T., Mashimo, D., Jamalipour, A., Hashimoto, K., Nemoto, Y., Kato, N.: ELB: an explicit load balancing routing protocol for multi-hop NGEO satellite constellations, pp. 1–5. IEEE Globecom, San Francisco (2006)
3. Taleb, T., Mashimo, D., Jamalipour, A., Kato, N., Ne-moto, Y.: Explicit load balancing technique for NGEO satellite IP networks with on-board processing capabilities. IEEE/ACM Trans. Netw. **17**(1), 281–293 (2009)
4. Taleb, T., Fadlullah, Z.M., Takahashi, T., Wang, R., Nemoto, Y., Kato, N.: Tailoring ELB for multi-layered satellite networks. In: IEEE International Conference on Communications, Dresden, pp. 1–5 (2009)
5. Yong, L., Sun, F., Zhao, Y., Li, H., Liu, H.: Distributed traffic balancing routing for LEO satellite networks. Int. J. Comput. Netw. Inf. Secur. **6**(1), 19–25 (2013)
6. Song, G., Chao, M., Yang, B., Zheng, Y.: TLR: a traffic-light-based intelligent routing strategy for NGEO satellite IP networks. IEEE Trans. Wirel. Commun. **13**(6), 3380–3393 (2014)
7. Zihe, G., Qing, G., Zhenyu, N., Ying, T., Naijin, L.: A distributed routing algorithm with traffic prediction in LEO satellite networks. Inf. Technol. **10**(2), 285–292 (2011)
8. Yuan, R.A.O., Ru-chuan, W.A.N.G., Xiao-long, X.U.: Load balancing routing for single-layered satellite networks. J. China Univ. Posts Telecommun. **17**(2), 92–99 (2010)
9. Rao, Y., Wang, R.-c.: QoS routing based on mobile agent for LEO satellite IP networks. J. China Univ. Posts Telecommun. **16**(6), 57–63 (2009)
10. Rao, Y., Wang, R.-c.: Agent-based load balancing routing for LEO satellite networks. Comput. Netw. **54**(17), 3187–3195 (2010)
11. Rao, Y., Zhu, J., Yuan, C.-a., Jiang, Z.-h., Lei-yang, F., Shao, X., Wang, R.-c.: Agent-based multi-service routing for polar-orbit LEO broadband satellite networks. Ad Hoc Netw. **13** (Part B), 575–597 (2014)
12. Houtian, W., Qi, Z., Xiangjun, X., Ying, T., Naijin, L.: Cross-layer design and ant-colony optimization based routing algorithm for low earth orbit satellite networks. China Commun. **10**(10), 37–46 (2013)

Multiuser Detection for LTE Random Access in MTC-Oriented Space Information Networks

Qiwei Wang, Guangliang Ren$^{(\boxtimes)}$, and Jueying Wu

School of Telecommunications, Xidian University,
No. 2 South Taibai Road, Xi'an 710071, China
qwwang@xidian.edu.cn,glren@mail.xidian.edu.cn

Abstract. The random access (RA) procedure in LTE systems is essential for the machine-type communication (MTC) in space information networks (SIN). By using the RA procedure, the delays of different MTC terminals could be aligned to the reference timing such that signals of each other could arrive at the satellite simultaneously. To this end, an enhanced iterative parallel interference cancellation (EIPIC) algorithm is proposed in this paper. Analyses show that the inverse impact of multiple access interference and near-far effect always persists even though the codes are ideally orthogonal with each other. The proposed algorithm could parallelly detect each possible code, re-construct its frequency-domain signal and cancel its inverse impact for further detection. And an optimized weighting vector is further applied for the frequency-domain signal re-construction based on the analyses of the mean square error. Simulation results show that the proposed algorithm could significantly improve the RA performance compared to existing algorithms.

Keywords: Long Term Evolution · Machine Type Communication · Space Information Network · Random Access · Multiuser detection

1 Introduction

The concept of space information networks (SIN) has been proposed for decades to provide global connectivity among any user terminals (UTs) that are separately located anywhere all over the world. In order to offer higher bandwidth and faster data rate, the SIN utilizes the most advanced civil communication systems as well, i.e., the long term evolution (LTE) system for satellite communications.

In LTE uplink, a random access (RA) procedure is designated for UTs to establish uplink synchronization [1], which is of great significance for MTC terminals that depend on the SIN. During the RA procedure, each UT randomly sends a Zadoff-Chu (ZC) code as a preamble through the physical random access

This work was supported in part by the National Natural Science Foundation of China (91538105), the 973 Program of China (2014CB340206), the China Postdoctoral Science Foundation (2016M590924), and the AeroSpace T.T.&.C. Innovation Program (201504A).

© Springer Nature Singapore Pte Ltd. 2017
Q. Yu (Ed.): SINC 2016, CCIS 688, pp. 104–116, 2017.
DOI: 10.1007/978-981-10-4403-8_10

channel (PRACH), and the satellite is required to detect active preambles and estimate corresponding parameters. The PRACH utilizes the single carrier frequency division multiple access (SC-FDMA) structure [1], by which a ZC code is generated in the transform-domain (TrD)[2,3] and transferred into the time-domain (TD) for transmission.

As the LTE system for SIN has many common points with the terrestrial LTE system, several researches have been done for the LTE RA procedure to accomplish the multiuser detection and estimation [4–10]. In [4], after the SC-FDMA demodulation at the base station (BS), a single user detector (SUD) is investigated by correlating the received TrD signal with each local ZC code while treating other codes as interference. If a peak correlation value exceeds a pre-defined threshold, the code is deemed as active. It is shown in [4] that the TrD correlation of the SUD could be equivalently realized by operating an inverse discrete Fourier transform (IDFT), and the round-trip delay (RTD) estimation in the LTE super coverage is investigated in [6,7]. As the threshold for the SUD is obtained empirically, an adaptive one is proposed in [8] to make it a constant false-alarm rate detector. An enhanced PRACH structure is proposed in [9], which directly generates ZC codes in the TD without SC-FDMA modulation but is not compliant with commercial LTE systems. A generalized likelihood ratio test (GLRT) algorithm is proposed in [10] by dividing the PRACH into several sub-channels in the frequency-domain (FD) and exploiting the frequency diversity gain among them. However, as the orthogonality of ZC codes is damaged, the GLRT performs even worse than the SUD algorithm. In available literature [3–8], although the TrD correlation is utilized for the multiuser detection, an explicit TrD signal model has not been presented yet, and none of them takes into account the multiple access interference (MAI) and near-far effect (NFE) which results because of imperfect power control [11], thus greatly degrading the multiuser detection performance.

Aiming at suppressing the MAI and NFE among multiuser signals, an iterative parallel interference cancellation (IPIC) based multiuser detection algorithm is proposed in [12]. On the basis of the space alternative generalized expectation-maximization (SAGE) algorithm [13], the IPIC algorithm is able to separate multiuser signals and mitigate interference among them by reconstructing and cancelling detected signals, thus significantly outperforming the conventional SUD algorithm. However, as the ZC codes are generated according to one or more ZC root sequences and their cyclic-shifted versions, the IPIC algorithm still suffers two drawbacks. For one thing, a large cyclic-shift region (which is much longer than the maximum timing delay) is necessitated to estimate the noise power and avoid interference between two neighboring codes (codes that are generated from the same root sequence with consecutive cyclic-shift regions), which introduces extra interference because a longer cyclic-shift region requires more ZC root sequences to generate an adequate number of codes and different ZC root sequences are not orthogonal with each other. For another, only the estimated TrD channel taps in the cyclic-shift region is utilized to reconstruct signals of detected codes, while the leakage energy of TrD channel taps is ignored.

In other words, the IPIC algorithm only mitigates interference among codes with different root values but not suppresses interference among codes that are cyclic-shifted with each other, indicating that the IPIC algorithm cannot fully mitigate interference of a detected code on others.

In this work, in order to completely mitigate interference of detected codes, an enhanced IPIC (EIPIC) algorithm is propose in this paper, and the main contributions are threefold.

- The MAI among possible codes is explicitly analyzed, by which we reveal that interference exists not only among codes with different roots, but also among code that are cyclic-shifted with each other due to the leakage energy of TrD channel taps.
- To mitigate the interference mentioned above, an EIPIC algorithm is proposed based on a TrD channel reconstruction technique, which involves the TrD channel estimation and its conversion from the TrD to TD. Based on a proper truncation of the TD channel for the current code, the TrD channel taps could then be reconstructed such that the interference of the current codes is able to be completely mitigated.
- When converting the channel estimation from the TrD to the TD, as the PRACH only occupies a narrow bandwidth, the conversion of the FD channel responses to the TD channel taps would lead to extra interference. In this case, an optimal FD weighting vector is further derived according the convex optimization theory.

This article is arranged as follows. Section 1.1 introduces the signal model. Section 2 illustrates the impact of MAI, proposes the EIPIC algorithm, and derives the FD weighting window. Simulation parameters and results are shown in Sect. 3, and the conclusion is presented in Sect. 4.

The notations are used as follows. Matrices and vectors are denoted by symbols in boldface, $\mathbf{X}^T$ and $\mathbf{X}^H$ denote the transpose and Hermitian of a matrix, respectively. $\mathbf{0}_{1 \times m}$ stands for a $1 \times m$ all-zeros vector. $\mathbf{1}_{1 \times m}$ stands for a $1 \times m$ all-ones vector. $\mathbf{X}[i_1 : i_2, j_1 : j_2]$ denotes a sub-matrix obtained by extracting rows i_1 through i_2 and columns j_1 through j_2 from a matrix $\mathbf{X}$. $\mathbf{x}((\tau))_M$ denotes a vector $\mathbf{x}$ having a cyclic shift by $(\tau \bmod M)$ points. $\lceil x \rceil$ is the smallest integer that is not smaller than x, and $\lfloor x \rfloor$ is the largest integer that is smaller than or equal to x.

1.1 Signal Model

In the PRACH [1] of SC-FDMA systems, the number of subcarriers is defined as N, among which M continuous ones are left for DFT pre-coded ZC codes. After downlink synchronization, each UE randomly selects a code from a ZC code set $\mathbb{C} = \{\mathbf{c}_1, ..., \mathbf{c}_V\}$ in the TrD, the vth code is given as $\mathbf{c}_v = [c_v(0), ..., c_v(M-1)]^T$, where $c_v(m)$ is the mth element of the vth ZC code.

Assume that the code $\mathbf{c}_{v_b}$ is selected by the bth UE among B active UEs, $1 \leqslant b \leqslant B < V$. An M-point DFT is applied to form the pre-coded code in the

FD as $\mathbf{C}_{v_b} = \mathcal{D}\{C_{v_b}(k), k=0,...,M-1\}$. Then $\mathbf{C}_{v_b}$ is mapped onto subcarriers with indexes of $\mathcal{J}_k = \mathcal{J}_0 + k$, where $\mathcal{J}_0$ is the starting index, and the TD signal is expressed as

$$s_{v_b}(n) = \frac{1}{\sqrt{N}} \sum_{k=0}^{M-1} C_{v_b}(k) e^{j2\pi \frac{\mathcal{J}_k}{N} n}, 0 \leqslant n \leqslant N - 1. \tag{1}$$

After adding the cyclic prefix and guard interval, the RA signals propagate through the multi-path channel and the TD received signal at the base station is given as

$$y(n) = \sum_{b=1}^{B} \sum_{l=d_{v_b}}^{d_{v_b}+L-1} h_{v_b}(l - d_{v_b}) s_{v_b}(n - l) + w(n), \tag{2}$$

where d_{v_b} is the round-trip delay (RTD), and $w(n)$ is the additive white Gaussian noise (AWGN). The maximum RTD is given as d_{max}, and the CIR has a maximum delay spread of $L - 1$ samples such that $h_{v_b}[l] = 0$ for $l < 0$ and $l \geqslant L$. After discarding the cyclic prefix and guard interval and defining a vector as $\mathbf{y} = [y(0),...,y(N-1)]^T$, the received signal in the FD is given by

$$\mathbf{Y} = \mathbf{Fy} = [Y(\mathcal{J}_0),...,Y(\mathcal{J}_{M-1})]^T = \sqrt{N} \cdot \sum_{b=1}^{B} \mathbf{C}_{v_b} \mathbf{Fh}_{v_b} + \mathbf{W} = \sum_{b=1}^{B} \mathbf{C}_{v_b} \mathbf{H}_{v_b} + \mathbf{W}, \tag{3}$$

where

- $\mathbf{F}$ is an $M \times N$ DFT matrix with its entry (k,n) represented as $\mathcal{F}(k,n) = \frac{1}{\sqrt{N}} e^{-j2\pi \frac{\mathcal{J}_k}{N} n}, k = 0, 1, ..., M - 1, n = 0, 1, ..., N - 1$;
- $\mathbf{h}_{v_b} = \left[\mathbf{0}_{1 \times d_{v_b}} \xi_{v_b} \mathbf{0}_{1 \times (N-L-d_{v_b})}\right]^T$ is defined as the equivalent TD CIR by incorporating the RTD, and $\xi_{v_b} = [h_{v_b}(0),...,h_{v_b}(L-1)]$ is the CIR of the current code;
- $\mathbf{H}_{v_b} = [H_{v_b}(\mathcal{J}_0),...,H_{v_b}(\mathcal{J}_{M-1})]^T = \sqrt{N} \cdot \mathbf{Fh}_{v_b}$ is the vector of channel frequency responses;
- $\mathbf{W} = \mathbf{Fw} = [W(\mathcal{J}_0),...,W(\mathcal{J}_{M-1})]^T$ is the AWGN in the FD with $\mathbf{w} = [w(0),...,w(N-1)]^T$.

Since ZC codes have been DFT-precoded and could preserve orthogonality in the TrD, the received FD signal in (3) should be transferred into the TrD. Defining $\mathbf{Q}$ as an M-point DFT matrix with its entry (k,m) given as $\mathcal{Q}(k,m) = \frac{1}{\sqrt{M}} e^{-j2\pi \frac{k}{M} m}, k = 0, 1, ..., M - 1, m = 0, 1, ..., M - 1$. As the TrD and FD are related by $\mathbf{Q}$, one can have [14]

$$\mathbf{Q}^H \mathbf{C}_{v_b} \mathbf{H}_{v_b} = \mathbf{c}_{v_b} \odot \mathbf{g}_{v_b} = \mathbf{\Omega}_{v_b} \mathbf{g}_{v_b}, \tag{4}$$

where $\mathbf{g}_{v_b} = \frac{1}{\sqrt{M}} \mathbf{Q}^H \mathbf{H}_{v_b} = [g_{v_b}(0),...,g_{v_b}(M-1)]^T$ is the TrD channel vector, $\odot$ denotes the cyclic convolution. $\mathbf{\Omega}_{v_b}$ is a Toeplitz matrix defined as $\mathbf{\Omega}_{v_b} = [\mathbf{c}_{v_b}((0))_M \quad \cdots \quad \mathbf{c}_{v_b}((M-1))_M]$, where $\mathbf{c}_{v_b}((m))_M$ denotes the vector $\mathbf{c}_{v_b}$ cyclic shifted by $(m \bmod M)$ points. $\mathbf{G} \triangleq \mathbf{Q}^H \mathbf{F}$ is defined as an $M \times N$ transfer matrix from the TD to TrD with its entry $\mathcal{G}(m, n)$ given as

$$\mathcal{G}(m, n) \triangleq \sum_{k=0}^{M-1} \mathcal{F}(k, n) \mathcal{Q}^\dagger(k, m) = \frac{1}{\sqrt{MN}} e^{-j\frac{2\pi \mathcal{J}_0}{N} n} \chi\left(m - \frac{M}{N} n\right), \quad (5)$$

and

$$\chi(m) = e^{j\pi m \frac{M-1}{M}} \frac{\sin \pi m}{\sin \frac{\pi m}{M}} = \frac{1}{M} e^{j\pi m \frac{M-1}{M}} \frac{sinc\,(\pi m)}{sinc\,\frac{\pi m}{M}}. \quad (6)$$

The received TrD signal is then represented in the matrix form as

$$\mathbf{r} = \mathbf{Q}^H \mathbf{Y} = \sum_{b=1}^{B} \mathbf{\Omega}_{v_b} \mathbf{g}_{v_b} + \varpi, \quad (7)$$

where $\varpi = \mathbf{Q}^H \mathbf{W} = [\varpi(0), ..., \varpi(M-1)]^T$ is defined as the AWGN in TrD.

2   Enhanced Iterative Parallel Interference Cancellation

In this section, an MAI analysis in the TrD is first given and the EIPIC based MUD algorithm is proposed by deriving a novel channel reconstruction technique, for which an optimized weighting window is derived based on the MMSE criterion.

Analysis of Multiple Access Interference. The correlation of the received TrD signal with a local code is expressed as

$$\Re\left(\mathbf{c}_v((m))_M \mid \mathbf{r}\right) = \mathbf{c}_v^H((m))_M \cdot \mathbf{r}, \ 0 \leqslant m \leqslant M - 1. \quad (8)$$

As the ZC code set is generated by several root ZC sequences and their cyclic-shift versions, it is assumed that the vth code is generated by the root sequence with a root value u and is selected by the first UE, i.e., $\mathbf{c}_{v_1} = \mathbf{s}_{u, p_1}((0))_M = \mathbf{s}_u((p_1 \cdot N_{cs}))_M$, where $\mathbf{s}_u((0))_M$ is a root ZC sequence with a root value u, N_{cs} is the length of the cyclic-shift region and $0 \leqslant p_1 < \lfloor M/N_{cs} \rfloor - 1$. The codes of the first β UEs are generated by the same root with $p_b \neq p_1, b = 2, ..., \beta$, and the other $B - \beta$ codes with a root ζ. The correlation is expressed as

$$\Re\left(\mathbf{c}_v((m))_M \mid \mathbf{r}\right) = \Re\left(\mathbf{s}_{u, p_1}((m))_M \mid \mathbf{r}\right) = \mathbf{s}_{u, p_1}^H((m))_M \mathbf{\Omega}_{u, p_1} \mathbf{g}_{u, p_1}$$

$$+ \mathbf{s}_{u, p_1}^H((m))_M \varpi + \mathbf{s}_{u, p_1}^H((m))_M \sum_{b=2}^{\beta} \mathbf{\Omega}_{u, p_b} \mathbf{g}_{u, p_b} + \mathbf{s}_{u, p_1}^H((m))_M \sum_{b=\beta+1}^{B} \mathbf{\Omega}_{\zeta, p_b} \mathbf{g}_{\zeta, p_b} \quad (9)$$

The first item in (2) is an element of the TrD channel estimates, the second item is a linear combination of the TrD AWGN, and the third and fourth items contribute to the MAI.

Since the transfer function $\mathcal{G}(m,n)$ of TD to TrD is Sinc-like, two conclusions could be derived below. One relates to the first item, which indicates that the energy of TrD channel taps not only concentrates into the region of $m \in [(-d, \tau_{\max} + d) \mod M]$, but also disperses in $m \in [\tau_{\max} + d + 1, M - d - 1]$, where d is defined as an extra window in the TrD, and $\tau_{\max} \triangleq \lceil (d_{\max} + L - 1)M/N \rceil$ is the corresponding maximum delay spread in the TrD [15]. The other one is that the MAI actually performs in two-fold. For one thing, the third item represents the MAI among ZC codes generated from the same root, which is caused by the energy leakage of TrD channel taps. For another, the last item represents the MAI among codes with different roots, which is caused by non-orthogonality of codes with different roots.

As for the SIC and IPIC algorithms, only the channel estimates in $m \in [(-d, \tau_{\max} + d) \mod M]$ could be obtained, indicating that the existing algorithms only mitigate the MAI caused by the forth item, i.e., MAI introduced by codes with different roots. In this case, the EIPIC algorithm is proposed to exploit the leakage energy of each code so that the MAI among codes not only with the same root but also with different roots could be both mitigated.

Enhanced Iterative Parallel Interference Cancellation. In this section, the proposed EIPIC algorithm would be proposed by introducing a channel reconstruction technique in accordance with the convex optimization theory.

(1) Re-expression of the IPIC algorithm

In [14] dealing with RA with frequency offsets, the IPIC algorithm detects the active codes by iteratively cancelling the impacts of other ones in the TD using frequency offsets compensation. However, with negligible frequency offsets, the IPIC algorithm could be re-expressed as follows.

At the ith iteration, the code indexes are re-ordered by re-arranging previous detection metrics of all possible codes from strong to weak, and the order is denoted by $\mathcal{K} = \{\kappa_p^{(i)}, p = 1, 2, ..., V\}$. The re-constructed FD signal of each code is defined as $\widehat{\mathbf{R}}_{\kappa_p^{(i)}}^{(i)}$. Initializing the iteration number $i = 1$, and $\widehat{\mathbf{R}}_{\kappa_p^{(1)}}^{(0)} = \mathbf{0}_{M \times 1}$, the IPIC algorithm is implemented as follows.

– **E-Step**

$$\widehat{\mathbf{Y}}_{\kappa_p^{(i)}}^{(i)} = \mathbf{Y} - \left(\sum_{q=1}^{p-1} \widehat{\mathbf{R}}_{\kappa_q^{(i)}}^{(i)} + \sum_{q=p+1}^{V} \widehat{\mathbf{R}}_{\kappa_q^{(i)}}^{(i-1)} \right) = \mathbf{R}_{\kappa_p^{(i)}} + \eta_{\kappa_p^{(i)}}^{(i)}, \qquad (10)$$

where

$$\eta_{\kappa_p^{(i)}}^{(i)} = \sum_{q=1}^{p-1} \left[\mathbf{R}_{\kappa_q^{(i)}} - \widehat{\mathbf{R}}_{\kappa_q^{(i)}}^{(i)} \right] + \sum_{q=p+1}^{V} \left[\mathbf{R}_{\kappa_q^{(i)}} - \widehat{\mathbf{R}}_{\kappa_q^{(i)}}^{(i-1)} \right] + \mathbf{W}. \qquad (11)$$

– **M-Step**

$$\widehat{\mathbf{R}}_{\kappa_p^{(i)}}^{(i)} = argmin \left\| \widehat{\mathbf{Y}}_{\kappa_p^{(i)}}^{(i)} - \mathbf{C}_{\kappa_p^{(i)}} \mathbf{H}_{\kappa_p^{(i)}} \right\|^2. \qquad (12)$$

To solve the M-Step above, a vector of the TrD channel estimation could be estimated as

$$\widehat{\mathbf{g}}_{\kappa_p^{(i)}}^{(i)} = \frac{1}{\sqrt{M}} \cdot \mathbf{Q}^H \mathbf{C}_{\kappa_p^{(i)}}^H \mathbf{Y}_{\kappa_p^{(i)}}^{(i)} = \frac{1}{\sqrt{M}} \mathbf{\Omega}_{\kappa_p^{(i)}}^H \mathbf{r}. \tag{13}$$

The variance of noise and interference can be estimated as

$$\widehat{\sigma}_{\kappa_p^{(i)}}^{(i)2} = \frac{1}{N_{cs} - \tau_{\max} - 2d} \sum_{m=\tau_{\max}+d}^{N_{cs}-1-d} \left| \widehat{g}_{\kappa_p^{(i)}}^{(i)}(m) \right|^2, \tag{14}$$

where $N_{cs} - 1 - d > \tau_{\max} + d$ should be guaranteed according to [15]. The detection metric is

$$\Psi_{\kappa_p^{(i)}}^{(i)} = \max_{m \in [(-d, \tau_{\max}) \bmod M]} \frac{1}{\widehat{\sigma}_{\kappa_p^{(i)}}^{(i)}} \left| \widehat{g}_{\kappa_p^{(i)}}^{(i)}(m) \right|. \tag{15}$$

Since the FD and TrD are related by $\mathbf{Q}$, the FD signal of the current code is reconstructed as

$$\widehat{\mathbf{R}}_{\kappa_p^{(i)}}^{(i)} = \mathbf{C}_{\kappa_p^{(i)}} \widehat{\mathbf{H}}_{\mathrm{IPIC},\kappa_p^{(i)}}^{(i)} = \mathbf{C}_{\kappa_p^{(i)}} \mathbf{Q}\mathbf{\Pi}\widehat{\mathbf{g}}_{\kappa_p^{(i)}}^{(i)}, if \ \Psi_{\kappa_p^{(i)}}^{(i)} \geq \lambda;$$

$$\widehat{\mathbf{R}}_{\kappa_p^{(i)}}^{(i)} = \mathbf{0}_{M \times 1}, if \ \Psi_{\kappa_p^{(i)}}^{(i)} < \lambda. \tag{16}$$

where $\mathbf{\Pi} = \mathcal{D}\{\mathbf{1}_{1\times(\tau_{\max}+d)} \ \mathbf{0}_{1\times(M-\tau_{\max}-2d)} \ \mathbf{1}_{1\times d}\}$ is a truncating matrix to extract the TrD channel estimates of the current code, λ is a pre-defined threshold, $\widehat{\mathbf{H}}_{\mathrm{IPIC},\kappa_p^{(i)}}^{(i)} = \mathbf{Q}\mathbf{\Pi}\widehat{\mathbf{g}}_{\kappa_p^{(i)}}^{(i)}$ is the reconstructed FD channel responses for the IPIC algorithm. Note that with the truncating matrix, the IPIC algorithm only obtain the TrD channel estimation in the current cyclic-shift region to re-construct the FD channel responses. However, as the energy of TrD channel taps spread over the whole TrD, whose length is far greater than that of the cyclic-shift region, the energy leakage cannot be ignored and it still leads to extra interference.

(2) Channel Reconstruction Technique

To further improve the performance of the IPIC algorithm, an channel reconstruction technique is derived to reserve as much channel information as possible for signal reconstruction and MAI mitigation. The basic idea is to obtain the CIR in the TD, which should reside in the TD region $n \in [(-d', d_{\max}+d'+L-1) \bmod N]$, where $d' = d\lceil N/M \rceil$ denotes the extra window in the TD corresponding to d in the TrD, $d_{\max}+d'+L-1 < (N_{cs}-1)\lceil N/M \rceil$. By transferring the TrD channel estimation into the TD and implementing another truncating operation, the reconstructed FD channel estimation for the EIPIC algorithm is represented as

$$\widehat{\mathbf{H}}_{\mathrm{EIPIC},\kappa_p^{(i)}}^{\prime(i)} = \mathbf{F}\mathbf{\Gamma}\mathbf{G}^H \widehat{\mathbf{g}}_{\kappa_p^{(i)}}^{(i)} = \mathbf{F}\mathbf{\Gamma}\mathbf{F}^H \mathbf{C}_{\kappa_p^{(i)}}^H \mathbf{Y}_{\kappa_p^{(i)}}^{(i)} \tag{17}$$

where $\mathbf{\Gamma} = \mathcal{D}\{\mathbf{1}_{1\times(d_{\max}+d'+L-1)} \mathbf{0}_{1\times(N-d_{\max}-2d'-L+1)} \mathbf{1}_{1\times d'}\}$ is the TD truncation to carve out the CIR of the current code. However, as the PRACH only occupies a narrow bandwidth, the DFT matrix $\mathbf{F}$ is an $M \times N$ matrix such that the conversion of the FD channel responses to the CIR would lead to the energy leakage to the CIR, indicating that the TD truncation would still leads to a loss of channel information. In this case, an optimized weighting window $\mathbf{a} = [a(0), ..., a(M-1)]^T$ is utilized on $\widehat{\mathbf{H}}'^{(i)}_{\text{EIPIC},\kappa_p^{(i)}}$ to mitigate the energy leakage in the TD and the re-constructed FD channel for the EIPIC algorithm is given by [16]

$$\widehat{\mathbf{H}}^{(i)}_{\text{EIPIC},\kappa_p^{(i)}} = \mathcal{D}^{-1}(\mathbf{a}) \, \mathbf{F}\mathbf{\Gamma}\mathbf{F}^H \mathcal{D}(\mathbf{a}) \, \mathbf{C}^H_{\kappa_p^{(i)}} \mathbf{Y}^{(i)}_{\kappa_p^{(i)}}. \tag{18}$$

(3) Derivation of the Optimized Weighting Window

The weighting window is derived by minimizing the MSE between the reconstructed FD channel responses with the real one, i.e., $\mathbf{H}_{\kappa_p^{(i)}}$. At the ith iteration, the MSE is represented as

$$\mathbf{MSE}_{\widehat{\mathbf{H}}^{(i)}_{\text{EIPIC}}}[\mathbf{a}] = \mathbf{E}\left(\sum_{p=1}^{V} \left\|\mathbf{H}_{\kappa_p^{(i)}} - \widehat{\mathbf{H}}^{(i)}_{\text{EIPIC},\kappa_p^{(i)}}\right\|^2\right) \tag{19}$$

Substituting (18) into (19), we have

$$\mathbf{MSE}_{\widehat{\mathbf{H}}^{(i)}_{\text{EIPIC}}}[\mathbf{a}] = \mathbf{E}\left(\sum_{p=1}^{V} \left\|\mathcal{D}^{-1}(\mathbf{a})\mathbf{F}\tilde{\mathbf{\Gamma}}\mathbf{F}^H \mathcal{D}(\mathbf{a})\mathbf{H}_{\kappa_p^{(i)}}\right\|^2\right)$$
$$+ \mathbf{E}\left(\sum_{p=1}^{V} \left\|\mathcal{D}^{-1}(\mathbf{a})\mathbf{F}\mathbf{\Gamma}\mathbf{F}^H \mathcal{D}(\mathbf{a})\mathbf{u}_{\kappa_p^{(i)}}\right\|^2\right) \tag{20}$$

where $\tilde{\mathbf{\Gamma}} = \mathbf{I} - \mathbf{\Gamma}$, and $\mathbf{u}_{\kappa_p^{(i)}} = \sum_{\iota \neq p}^{V}\left(\mathbf{C}^H_{\kappa_p^{(i)}}\mathbf{C}_{\kappa_\iota^{(i)}}\mathbf{H}_{\kappa_\iota^{(i)}}\right) + \mathbf{W}$. The former item in (20) is the energy leakage of the current CIR that lies in $n \in \{(L+d_{\max}+d' : N-d'-1)\}$ in the TD, representing a channel-related term (CRT). The latter one represents the leakage energy of codes that impacts the current one along with the AWGN, denoting a noise-related term (NRT)[16].

Since the MSE consists of the CRT and NRT, it is supposed to suppress both of them. However, a direct minimization may lead $\mathcal{D}(\mathbf{a})$ ill-conditioned so that the NRT would be amplified. To this end, it is supposed to minimize the CRT while suppressing the amplification of the NRT, and the following optimization problem could be obtained according to [16] as

$$\max_{\mathbf{a}} \left|\mathbf{1}^T\mathbf{a}\right|$$

$$s.t. \left\|\mathbf{F}\tilde{\mathbf{\Gamma}}\mathbf{F}^H \mathcal{D}(\phi^n)\mathbf{a}\right\|^2 \leq \gamma \left\|\mathbf{F}\tilde{\mathbf{\Gamma}}\mathbf{F}^H \mathcal{D}(\phi^n)\mathbf{1}\right\|^2, \Im\left(\mathbf{1}^T\mathbf{a}\right) = 0, \|\mathbf{a}\|^2 \leq M \tag{21}$$

where $\phi^n = \left[e^{\frac{-j2\pi n J_0}{N}}, ..., e^{\frac{-j2\pi n J_0}{N}}\right]^T$, γ is the CRT suppression factor, The problem (21) is the second-order cone programming and can be solved with optimization tools. The optimized weighting window could be derived offline, bringing no much computational burden. The reconstructed FD signal of the current code is finally given as $\widehat{\mathbf{R}}_{\kappa_p^{(i)}}^{(i)} = \mathbf{C}_{\kappa_p^{(i)}} \widehat{\mathbf{H}}_{\text{EIPIC},\kappa_p^{(i)}}^{(i)}$ when $\Psi_{\kappa_p^{(i)}}^{(i)} \geq \lambda$.

3.4 Further Discussion of PRACH in SIN

Although LTE systems could be utilized into the SIN or satellite communications, there are still several issues that need to be discussed in detail, e.g., the PRACH structure for the SIN, the power consumption and transmission delays of MTC terminals, etc.

For one thing, as it is known that the distance of a terrestrial terminal from the satellite is much longer than that from a terrestrial base station, the PRACH structure should be modified such that the duration of the PRACH slot τ_{PRACH} be required to be longer than the maximum RTD d_{max}, i.e., $\tau_{PRACH} > d_{RTD,max}$. However, as the height of the satellite is known to the communication system, by assuming the coverage of a satellite's beam is a circle, the maximum RTD could be then divided into two items, i.e., $d_{RTD,max} = d_{satellite} + \Delta d$, where $d_{satellite}$ indicates the RTD from a terminal located at the origin of the circle to the satellite, and Δd is the RTD difference of terminals between other locations and the origin point of the circle. Therefore, the LTE system for the satellite is only required to estimate the RTD difference of different M2M terminals, while $d_{satellite}$ could be pre-compensated by the timing advance mechanism. As Δd is relatively much smaller than d_{RTD}, the existing PRACH format-3 or the modified S-LTE PRACH structure are assumed to be sufficient to accomplish the random access procedure of LTE systems for the SIN.

For another, the power consumption issue is also a critical problem for the MTC terminals since the battery of the MTC terminals is usually limited and non-chargeable. In this case, the power of MTC terminals is not possible to be large enough for satellite communications. Aiming at dealing with this issue, existing literature [18–20] has been proposed by either using an ad-hoc network with main cluster heads or several terrestrial amplify-and-forward (AF) stations. By doing so, signals of MTC terminal would be amplified by active amplifiers such that satellites could receive PRACH signals with SNR high enough and the power consumption of M2M terminals is reduced to the minimum.

Furthermore, in Sect. 3, by considering these practical issues, both the channel model and RTD of M2M terminals would be explicitly analyzed and configured.

3   Simulation Results

Figures 1 shows the comparison of miss-detection performance between the EIPIC algorithms with and without the optimized weighting window. It is seen that the EIPIC performs better with the optimized window, indicating that the derived window could improve the MUD performance so that the capacity of RA is improved as well. Note that performance of all the algorithms with SNR

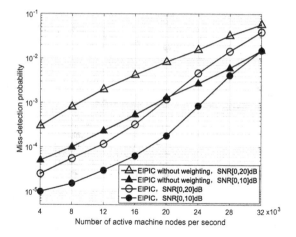

Fig. 1. Comparison of miss-detection between the EIPIC algorithms with and without the optimization weighting vector.

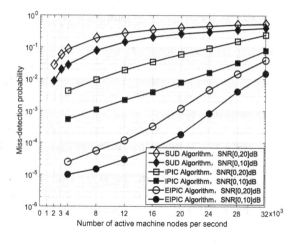

Fig. 2. Performance of miss-detection among the SUD, IPIC, and EIPIC.

in $[0, 20]$dB is worse than that with SNR in $[0, 10]$dB, verifying that the MUD performance is mainly affected by the MAI and the NFE rather than the AWGN.

Figures 2 shows the comparison of miss-detection performance among the SUD, IPIC, and EIPIC algorithms. It is obviously seen that the proposed EIPIC algorithm provides the best detection performance. The miss-detection probability of the EIPIC decreases three orders of magnitude compared to the SUD and two orders of magnitude compared to the IPIC with SNR in $[0, 20]$dB when the number of active MTC devices is less than 20×10^3 per second, indicating that the EIPIC could significantly improve the detection performance. With further increase in the number of active MTC devices, the improvement is reduced but

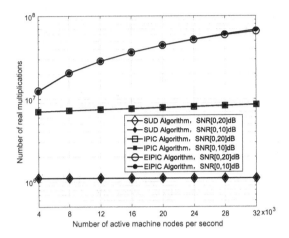

Fig. 3. Comparison of complexity among the SUD, IPIC, and EIPIC.

still exists. According to the performance specification [17], the missed detection probability should not be larger than 1%. In this case, it can be seen that the EIPIC can accommodate approximate 25 times as many MTC devices as the SUD and 3 times as many as the IPIC with SNR in $[0, 20]$dB. With SNR in $[0, 10]$ dB, the number of accommodated MTC devices of the EIPIC is 15 times that of the SUD and 1.5 times that of IPIC, showing that the capacity of RA procedure could be greatly improved.

Figure 3 shows the comparison of complexities among these algorithms. Since a real multiplication costs much more in hardware than a real addition, we evaluate the complexity in terms of real multiplications. We see that the EIPIC algorithm requires more complexity than the IPIC algorithm as the number of active MTC devices increases. However, the extra complexity is paid off to greatly increase both the performance and capacity of RA procedure.

4   Conclusion

In this paper, we show that the MAI and NFE always exist so as to degrade the RA performance even though the ZC codes are ideally orthogonal with each other. By analysing the shortages of the IPIC algorithm, an EIPIC with weighting optimization is proposed to significantly improve the RA performance, where the design of weighting vector is based on the purpose of minimizing the MSE of the estimation of the FD channel of each valid code, leading to better service quality and user experience of wireless networks.

References

1. 3GPP, Evolved Unversal Terrestrial Radio Access (E-UTRA), User Equipment radio transmission and reception, 3Gpp. Technical Specification 36.211, v.11.4.0, February 2013
2. Yang, L., Ren, G., Yang, B., Qiu, Z.: Fast time-varying channel estimation technique for LTE uplink in HST environment. IEEE Trans. Veh. Tech. **61**(9), 4009–4019 (2012)
3. Mansour, M.M.: Optimized architecture for computing Zadoff-Chu sequences with application to LTE. In: 2009 Global Telecommunication Conference (GLOBECOM 2009), pp. 1–6, November 2009
4. Sesia, S., Toufik, I., Baker, M.P.J.: The UMTS Long Term Evolution - From Theory to Practice. Wiley, New York (2009)
5. Yu, C., Xiangming, W., Wei, Z.: Random access algorithm of LTE TDD system based on frequency domain detection. In: Fifth International Conference on Semantics, Knowledge and Grid (2009)
6. Kim, S., Joo, K., Lim, Y.: A delay-robust random access preamble detection algorithm for LTE system. In: IEEE Radio and Wireless Symposium (RWS), pp. 15–18, January 2012
7. Wei, L., Li, X., Yaying, W., et al.: Time-domain-cascade-correlation timing advance estimation method in LTE-A super coverage. In: International Conference on Wireless Communication and Signal Processing (WCSP), pp. 1–6, October 2013
8. Yanchao, H., Han, J., Tang, S.: A method of PRACH detection threshold setting in LTE TDD femtocell system. In: Conference on Communication and Networking in China (CHINACOM), pp. 408–413, August 2012
9. Yang, X., Fapojuwo, A.O.: Enhanced preamble detection for PRACH in LTE. In: IEEE Wireless Communications and Networking Conference (WCNC), pp. 3306–3311, April 2013
10. Sanguinetti, L., Morelli, M., Marchetti, L.: A random access algorithm for LTE systems. Trans. Emerg. Telecommun. Tech. **26**, 3306–3311 (2013)
11. Yang, Y., Yum, T.S.P.: Analysis of power ramping schemes for UTRA-FDD random access channel. IEEE Trans. Wirel. Commun. **4**(6), 2688–2693 (2005)
12. Andrews, J.G., Meng, T.H.-Y.: Performance of multicarrier CDMA with successive interference cancellation in a multipath fading channel. IEEE Trans. Commun. **52**(5), 811–822 (2004)
13. Fang, L., Milstein, L.B.: Successive interference cancellation in multicarrier DS/CDMA. IEEE Trans. Commun. **48**(9), 1530–1540 (2000)
14. Cheng, S., Narasimhan, R.: Soft interference cancellation receiver for SC-FDMA uplink in LTE. In: IEEE wireless Communications and Networking Conference (WCNC), pp. 3318–3322 (2013)
15. Zhang, H., Li, Y., Yi, Y.-W.: Practical considerations on channel estimation for up-link MC-CDMA systems. IEEE Trans. Wirel. Commun. **7**(1), 4384–4392 (2008)
16. Niridakis, N.I., Vergados, D.D.: A survey on the successive interference cancellation performance for single-antenna and multiple-antenna OFDM systems. IEEE Commun. Surv. Tutor. **15**(1), 312–335 (2013)
17. Benvenuto, N., Carnevale, G., Tomasin, S.: MC-CDMA with SIC: power control by discrete stochastic approximation and comparison with OFDMA. In: IEEE International Conference on Communications, pp. 5715–5720, June 2006
18. Sanctis, M.D., Cianca, E., Araniti, G.: Satellite communications supporting internet of remote things. IEEE Internet Things J. **3**(1), 3340–3350 (2016)

19. Liu, G., Tan, R., Zhou, R.: Volcanic earthquake timing using wireless sensor networks. In: Information Processing in Sensor Networks, pp. 91–102 (2013)
20. Kawamoto, Y., Nishiyama, H., Kato, N.: A centralized multiple access scheme for data gathering in satellite-routed sensor system (SRSS). In: IEEE Global Communications Conference (GLOBECOM), pp. 2998–3002 (2013)

Research Progresses and Trends of Onboard Switching for Satellite Communications

Wanli Chen[1,2,3], Kai Liu[4], and Xiang Chen[1,2,3(✉)]

[1] School of Electronics and Information Technology,
Sun Yat-Sen University, Guangzhou, China
[2] SYSU-CMU Shunde International Joint Research Institute(JRI),
Shunde, Guangdong, China
Chenxiang@mail.sysu.edu.cn
[3] Key Lab of EDA, Research Institute of Tsinghua
University in Shenzhen, Shenzhen, China
[4] China Academy of Electronics and Information Technology, Beijing, China

Abstract. As broadband satellite communication (SAT-COM) systems develop rapidly, Onboard Switching (OBS) technologies with flexible connectivity has attracted much attention by academic researchers and industry. However, due to the threat by the space radiation effects and the firm constraints of payload processing and power resources, OBS technologies are facing to great challenges in terms of its reliability and scalability. In the past literatures, in order to improve the connectivity and reliability of the single-path Crossbar switch, the multi-path Clos network is widely investigated for preferable adoption in OBS. Nevertheless, due to the extensive use of centralized scheduling, the current Clos network with low scalability is not applicable for high-throughput OBS systems. In this paper, by analyzing progresses and trends of decades of related researches, the developing directions of OBS are provided and discussed, combining the switching architecture, the queuing strategy and the corresponding scheduling algorithms. By employing specific queuing structure to sup-port distributed scheduling in the Clos network and along with distributed, fault-tolerant scheduling algorithms, the requirements for high scalability and fault tolerance of the OBS systems can be efficiently guaranteed, which is expected to be a new trend of OBS researches.

Keywords: Onboard switching · Clos-network · In-sequence service · Distributed fault-tolerant scheduling

1   Introduction

Developing from 1960s, the SATCOM system is characterized by wide bandwidth, broad coverage, high flexibility, the independence from geographical conditions and the suitability for multiple traffics. Due to the rapidly developing traffics, the existing civil broadband satellite systems are taking higher and higher frequency bands instead of lower frequency bands, in order to obtain higher system capacities. Meanwhile, the services of SATCOM have also transformed from the single service into integrated broadband services including voice, data and various multimedia services.

© Springer Nature Singapore Pte Ltd. 2017
Q. Yu (Ed.): SINC 2016, CCIS 688, pp. 117–125, 2017.
DOI: 10.1007/978-981-10-4403-8_11

In [1], the transponder technologies in the SAT-COM systems can be divided into the bent-pipe transparent transponder and the developing "switch-in-the-sky". In a bent-pipe transponder, the satellite will amplify the signals and translate the frequency, without performing the signal detection, decoding or protocol translation. Lots of broadband SATCOM systems, e.g. Astrolink and SkyBridge, employ bent-pipe transponder relays. Well performed OBS is the core technology of "switch-in-the-sky". The end-to-end communication in OBS is accomplished within one satellite hop, resulting in reduced transmission delay only depending on ground station processing.

With the increasing rate of satellite links and the growing scale of satellite's beams, the onboard switching architecture is becoming the bottleneck of OBS. Actually, the onboard switching architecture suffers from the same problems as that of on-ground switching architecture, for example, processing complexity. Besides, the OBS architecture is firmly restricted by the constraints of payload processing and power resources, and also its reliability is threaten by the space radiation effects, e.g. single event effects. Interested readers can refer to [2] for the space radiation effects. Due to the risks of faults and invalidation in space environments, the capability of higher fault tolerance is necessary for future on-board switching architectures and scheduling algorithms. Therefore, in this paper, we will review the research progresses of OBS techniques, with special focuses on fault tolerance. Then some attractive trends of OBS are discussed and expected to be further investigated in the future.

The remainder of this article will be organized as follows. In Sect. 2, by analyzing the topology, queuing strategy and scheduling algorithms, we focus on the introduction of OBS architectures, including the Crossbar and the Clos network. In Sect. 3, induced by the requirements of the reliability and in-sequence guarantee in the switching system, the key research directions of future OBS are presented and analyzed. Finally, Sect. 4 concludes the paper.

2 Onboard Switching Architecture

2.1 Categories of OBS Architecture

With the increasing capacity of routers, the OBS fabric developed from Time-Division (TD) Switches to Space-Division (SD) Switches [3]. TD switches consist of shared memory and shared bus, as shown in Fig. 1. Shared memory is subject to the speed of reading and writing memories, while the shared bus is stringent with the bus bandwidth. Hence, TD switches are not applicable for the in-creasing requirements of switch size and port speed. Considering the uniqueness of switch path, the SD switches can be divided into two types: single-path structure such as Crossbar, and multi-path structure such as Clos network [3], with reference to Figs. 2 and 3. Due to the high reliability and the independence of each path, the multi-path SD switches are suitable for high-speed, high-throughput switches. Furthermore, as one of the multi-path SD Switch structures, the three-stage Clos network is more preferable and widely studied for its higher connectivity and greater reliability.

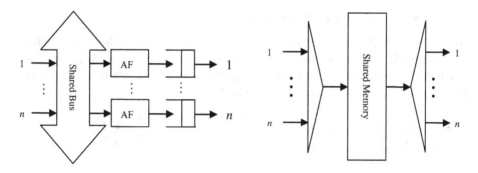

Fig. 1. Structure of shared bus and shared memory, (a) Shared Bus, (b) Shared Memory

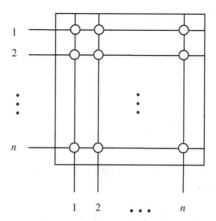

Fig. 2. Crossbar switch fabric

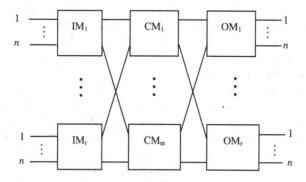

Fig. 3. A three-stage Clos network $C(n, m, r)$

2.2   Crossbar and Its Queuing Structures

As a single-path SD switch, the Crossbar switch is a basis of the multi-stage switches. Considering the buffer location of the Crossbar, the queuing structures can be classified into Input Queuing (IQ), Output Queuing (OQ), Virtual Output Queuing (VOQ), Crosspoint Queuing (CQ), etc. [4], as shown in Fig. 4.

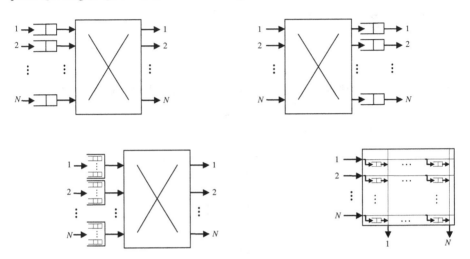

Fig. 4. Crossbar queuing structures, (a) Input queuing, (b) Output queuing, (c) Virtual output queuing, (d) Crosspoint queuing

OQ: In view of the Quality of Service (QoS), the OQ structure is supposed to perform the best. However, due to its current limited cache access rate, the OQ is not applicable for high-speed and high-throughput OBS systems.

IQ: In the IQ structure, there is a First In First Out (FIFO) queue allocated at each input port. If the traffic matrix of the system is given, the matrix can be processed by Birkhoff-von Neumann (BvN) decomposition. However, the traffic matrix is always unknown. In this case, the bipartite-graph based scheduling algorithms perform better, which include size-based matching algorithms, e.g. iRRM, DRRM and EDRRM, and weight-based matching algorithms, e.g. LQF, OCF, LPF. In [3] based on different properties, algorithms in IQ can be divided into maximum matching, maximal matching, randomized matching, frame-based matching and stable matching algorithms with speedup. As one of the disadvantages, adopting FIFO queuing strategy in IQ will result in the Head of Line (HoL) blocking phenomenon, which degrades the performance. Another disadvantage is that the scheduler in IQ works on a high frequency. In order to emulate the ideally optimal queuing, i.e. OQ, it needs to further in-crease the speedup of the switching structure. In this way, however, the traffic arrival rate at the output is higher than the link rate, causing an imbalance. From a global point of view, in order to solve the contention problem, it is inevitable to employ centralized scheduling algorithms, which leads to the scalability degradation of the system. Overall, the IQ structure is not suitable for the OBS.

VOQ: In order to eliminate the HoL blocking problem in IQ, Tamir proposed the VOQ mechanism in 1988 and later Anderson completed it in 1990s. As shown in Fig. 4c, there are N queues being set in each input port, which correspond to N output ports respectively. Originally with one queue in each input port, if the corresponding out-put port for the HoL cell is not available, then the HoL cell needs to wait for it to be available, and thus the following cells in the queue need to wait for the HoL cell to be transmitted. However, in VOQ, the aforementioned "following cells" can be transmitted immediately if their own corresponding output ports are available. Therefore, the HoL blocking problem is solved. At present, the VOQ status of the switch is widely used in the IQ mechanism.

CIOQ: In order to resolve the imbalance be-tween traffic arrival rate and the link rate in IQ, it is favorable to add an OQ at the output. In others words, the Combined Input and Output Queuing (CIOQ) structure is formed. Actually, just like the queuing strategy IQ, the CIOQ still shows the same deficiency of adopting centralized algorithms and scheduling with a high frequency. In [5], Iyer et al. showed how packets should be allocated in order to perform better and emulate an ideal OQ router. Algorithms to emulate the OQ are called stable matching algorithms, e.g. CCF, MUCFA and LOOFA.

CQ: Due to the mutual independence among the crosspoints, the CQ can fully resolve the output contention and reduce the time complexity.

CICQ: By combining the structure VOQ and CQ, the Combined Input and Crosspoint Queuing (CICQ) structure reduces the cache sets at the crosspoints, and thus improves the utilization of queuing resources. Various scheduling algorithms, e.g. LQF-RR, OCF-OCF, RR-RR [6], can be adopted in the CICQ mechanism.

The single-stage structure Crossbar is a basis of three-stage Clos network. To some extent, the current studies on the scheduling methods of Clos network are just the exten-sions of that of Crossbar, along with new problems for Clos network particularly.

2.3 Clos Network

With the background of circuit-switched (CS) network, C. Clos [7] first proposed the Clos net-work in 1950s. In the Clos network, the interconnection among multi-stage switching modules is conducive to producing a strictly non-blocking CS structure with less crosspoints. In the Clos network, the problems need to be considered include load balancing, multicast support and the overhead of interstage communication.

As shown in Fig. 3, the Clos network, $C(n, m, r)$, consists of r input modules (IMs) of size $n \times m$ in the input stage, m central modules (CMs) of size $r \times r$ in the middle stage, and r output modules (OMs) of size $m \times n$ in the output stage. In [8], the Clos network can be classified into two types: buffered and bufferless Clos network. Based on the internal buffer location in the net-work, the Clos network can be classified into five types. The qeuing structure, scheduling algorithms and existing problems of five types of Clos network are shown in Table 1.

Table 1. Five types of Clos network

Network Type	Queuing Structure	Scheduling Algorithm	Fault-tolerance	Existing Problems
SSS-Clos	IM: Bufferless CM: Bufferless OM: Bufferless	Euler partition, m_matching MAC schemes [9]	Matrix Decomposition Maximal Matching	Centralized scheduling
SSM-Clos	IM: Bufferless CM: Bufferless OM: XB	WMF-NP, WCMM		Centralized scheduling; Intra-competition of the network not alleviated;
MSM-Clos	IM: Shared VOQ CM: OQ OM: OQ	Weight-based: MWMD, DWMD [10]; Non-weight-based: Divide-and-conquer Dispatching Scheme, SRRD, CRRD [11]		Centralized scheduling; Competition of central stage unsettled; Too much control information to be switched, affecting the scheduling judge and the scalability; Need for complicated dispatching algorithm; OQ: Speedup limit
MMM-Clos	IM: VOQ CM: OQ OM: OQ	Adopting RR in the CM: MM^{e}M, TrueWay, MMM-IM, MMM-OM; Padded-frame based: EPF, FIM[3]; Heuristic matching: MCS, LDVSA [13]	Distributed Fault-toleration	Complicated design of queuing; OQ: Speedup limit
SMM-Clos	IM: Bufferless CM: OQ or VOQ OM: OQ or VOQ	MF-DSRR, MFRR		OQ: Speedup limit

Note that, among these five types of network, SMM-Clos network and MMM-Clos network are supportive to distributed scheduling. In [12], the designing principles for the OBS scheduling algorithm were proposed, one of which is adopting distributed scheduling. By adopting the distributed scheduling mechanism, each stage of switching module can be scheduled respectively and concurrently. As a result, the configuration time and the configuration complexity can be reduced and the scalability of switching can be increased. In this way, the efficiency of high-throughput OBS can be improved. Based on the above analysis, it can be indicated that SMM-Clos network and MMM-Clos network are suitable for the OBS.

3   Network Reliability and Capability of in-Sequence Guaranteed Scheduling

3.1   Network Reliability

Compared with the single-path switching architecture, the Clos network is more reliable for its higher connectivity. In the Clos network, in order to tolerate the faults brought by space radiation effects, Carpinelli *et al.* [14] proposed the network fault-tolerant Clos (FTC). In each stage of the FTC network, there is an additional switching module. By providing multi-path in each stage, the FTC improves the fault-tolerance after the central stage. Basing on the measurement of improving reliability in terrestrial switching, each node of the switching structure in the router Juniper T640 has an additional switch plane for backup. However, this kind of fault toleration method will further complicate the implementation and increase the payload of the satellite. In order to analyze the impact of interstage link faults, in [15], the authors proposed an analytical model on the blocking probability of a fault-tolerant network. Furthermore, in order to analyze the capability of fault tolerance when faults occur in each stage of the Clos network, Ref. [16] established a fault model on losing-contact faults in the network. In [12], by interconnecting planes, Liu *et al.* proposed the input and output parallel Clos (IOP-Clos) network and the interconnected parallel Clos (IP-Clos) network, both of which improve the performance of fault tolerance without increasing the complexity of the scheduling algorithm.

The Clos network outperforms the Crossbar for its multiple paths, high connectivity and high re-liability. Moreover, adopting distributed algorithms in the Clos network improves the performance of scalability. A higher scalability is preferable for a stable performance under the high-speed, high-throughput OBS system. Therefore, by adopting distributed algorithms in the Clos network, the whole system can be more stable and reliable.

3.2   In-Sequence Guarantee of Scheduling Algorithm

In the Clos network, as mentioned above, the SMM-Clos network and the MMM-Clos network are supportive to the distributed scheduling. However, both kinds of network suffer from the out-of-sequence (OOS) problem of switch queuing. Note that, when setting buffer for the central stage, cells from the same input are transferred with different path and thus with different delay for the output. This is the reason that causes the OOS problem of switch queuing. The OOS problem can be classified into two categories: in-packet and inter-packet OOS problems. The in-packet OOS problem is the case where cells from the same packets are out of sequence, while inter-packet OOS problem is the case where cells be-longing to different packets are out of sequence. Taking the MMM-Clos network for instance, Ref. [13] showed three methods to solve the above OOS problems:

(1) Controlling the OOS cases, e.g. the MMeM switching with rearranged output buffer;

(2) Avoiding the OOS conditions to occur, by forcing the cells belonging to the same flow to take the same switching path, e.g. the TrueWay switching with hash method;

(3) Maintaining the sequence order of transmit-ted cells with feedback mechanism, e.g. the MMM-IM with time-stamp monitoring mechanisms at the input modules of a switch.

When designing practical switching topologies in the OBS system, adopting the Clos network will outperform other competitive alternatives in terms of high reliability. Additionally, due to the need for fault tolerance, interconnected multiple planes should be employed. In this way, the connectivity and the reliability of the Clos network can be improved. When outlining a queuing strategy, it is supposed to take full account into the scalability of both the switching architectures and the scheduling algorithms. For in-stance, the emerging queuing fabric CICQ can be incorporated into the SMM- or MMM-Clos network to support the distributed scheduling. As for specific designing of scheduling algorithms, the aforementioned three methods for OOS problem, or the combination of them will be conducive to in-sequence guarantee of the scheduling.

Based on the above analysis, it can be predicted that, for uniform traffic with Bernoulli, by employing switch architecture similar to that in Ref. [12], it is feasible to design a fault-tolerant, in-sequence guaranteed algorithm to achieve a higher tolerance of faulty crosspoint than that of FDFT algorithm proposed in [12]. Meanwhile, the OOS problem can be avoided and 100% of throughput can be kept. As a result, an attractive performance of the whole OBS system can be achieved.

4   Conclusion and Future Work

In this paper, the existing OBS architectures, queuing strategies and scheduling algorithms in the broadband SATCOM systems are reviewed with detailed investigations. As the OBS is under the threat of space radiation effects, high capability of fault tolerance is required when designing the on-board switch. In addition, due to the in-creasing requirements of switch size and port speed, the scheduling algorithms in the OBS are supposed to be more scalable than that in the terrestrial switches. Therefore, the distributed scheduling in the Clos network, which produces higher scalability, will be preferable for the OBS. There are two kinds of Clos networks that support distributed scheduling, while result in the OOS problem. Although the existing rearrangement at the output port can solve the OOS problem, it is prohibitively costly to suit the OBS. There is hence a strong demand for a well-performed scheduling algorithm with in-sequence guarantee in the switch. In summary, in the OBS system, the design of an efficient queuing mechanism by distributed, fault-tolerant and in-sequence guaranteed scheduling is becoming an important re-search direction for onboard switching in the future.

References

1. Farserotu, J., Prasad, R.: A survey of future broadband multimedia satellite systems, issues and trends. IEEE Commun. Mag. **38**(6), 128–133 (2000)
2. Maurer, R.H., Fraeman, M.E., Martin, M.N., et al.: Harsh environments: space radiation environment, effects, and mitigation. Johns Hopkins APL Tech. Dig. **28**(1), 17–29 (2008)
3. Chao, H.J., Liu, B.: High performance switches and routers. Setit.rnu.tn **17**(2), 1009–1012 (2007)
4. Hu, H., et al.: Integrated uni- and multicast traffic scheduling in buffered crossbar switches. In: International Conference on Communications and Networking in China, Hangzhou, Zhejiang, China, 2008, Chinacom, pp. 66–72. IEEE (2008)
5. Iyer, S., Mckeown, N., et al.: On the speedup required for a parallel packet switch. IEEE/ACM Trans. Netw. **5**(6), 269–271 (2001)
6. Rojas-Cessa, R., Oki, E., Jing, Z., et al.: CIXB-1: combined input-one-cell-crosspoint buffered switch. In: 2001 IEEE Workshop on High Performance Switching and Routing, Dallas, USA, pp. 324–329 (2001)
7. Clos, C.: A study of non-blocking switch networks. Bell Syst. Tech. J. **32**(2), 406–424 (1953)
8. Gao, X., Qiu, Z., Zhang, M., et al.: Padded-frame based in-sequence dispatching scheme for Memory-Memory-Memory (MMM) Clos-network. J. Electron. Inf. Technol. **34**(11), 2715–2720 (2012)
9. Chao, H.J., Jing, Z., Liew, S.Y.: Matching algorithms for three-stage bufferless Clos network switches. IEEE Commun. Mag. **41**(10), 46–54 (2003)
10. Gao, Y., Qiu, Z., Zhang, M., et al.: Distributed weight matching dispatching scheme in MSM Clos-network packet switches. IEEE Commun. Lett. **17**(3), 580–583 (2013)
11. Oki, E., Jing, Z., Rojas-Cessa, R., et al.: Concurrent Round-Robin-Based dispatching schemes for Clos-network switches. IEEE/ACM Trans. Netw. **10**(6), 830–844 (2002)
12. Liu, K.: Research on key techniques for high speed onboard switching. Tsinghua University, Beijing (2015)
13. Gao, Y.: Study of Load-Balanced Scheduling Algorithms in Multi-stage Packet Switching Networks. Xidian University, Xi'an (2014)
14. Carpinelli, J.D., Nassar, H.: Fault-Tolerance For Switching Networks. Switching Networks: Recent Advances, pp. 1–23. Kluwer Academic Publishers, New York (2001)
15. Haynos, M.P., Yang, Y.: An analytical model on the blocking probability of a fault-tolerant network. IEEE Trans. Parallel Distrib. Syst. **10**(10), 1040–1051 (1999)
16. Yang, Y., Wang, J.: A fault-tolerant rearrangeable permutation network. IEEE Trans. Comput. **53**(4), 414–442 (2004)

Contact Graph Routing with Network Coding for LEO Satellite DTN Communications

Cuiqin Dai[1,2(✉)], Qingyang Song[1], Lei Guo[1], and Qianbin Chen[2]

[1] School of Computer Science and Engineering, Northeastern University,
Shenyang, People's Republic of China
{songqingyang,guolei}@mail.neu.edu.cn
[2] Key Lab of Mobile Communications Technology, Chongqing University of Posts
and Telecommunications, Chongqing, People's Republic of China
{daicq,chenqb}@cqupt.edu.cn

Abstract. In Delay/Disruption Tolerant Networks (DTNs), most previous works proposed to re-forward data and recalculate the route between the local node and one of its neighbors depending on the priority of message. However, some lower priority messages in a contact are replaced by the higher priority messages, which leads to the lower throughput and higher data loss rate. In this paper, we focus on the route selection process with the consideration of network coding in Low Earth Orbit (LEO) satellite DTN communications, and propose an improved Contact Graph Routing (CGR) scheme by introducing the Destination based Network Coding (DNC), namely DNC-CGR, to improve the network throughput and reduce the number of messages in the Inter-Satellite Links (ISLs). Simulation results show that the proposed DNC-CGR scheme can significantly improve network performance in comparison to the existing CGR schemes.

Keywords: DTN · CGR · NC · Satellite communications

1 Introduction

Delay-Disruption-Tolerant Network (DTN) emerged as a potential solution to cope with the challenges imposed by the occasionally-connected information transmission in the space environment including the satellite communication networks with the features of moderate delays, relatively low bandwidth and periodic connectivity [1, 2].

As an active research topic, the routing in DTN has been a very complex problem caused by the characteristic of intermittent connectivity and frequent network partition since the early 2000s. Some literatures introduced the concept of "contact" to the routing in DTN to convert the dynamic connection into the relative static connection. As early as 2004, Jain et al. [3] introduced a series of "contacts" based on the feature of intermittent connectivity to discuss the routing problem in DTNs. Each "contact" represented a period of time during which some data could be successfully forwarded from one DTN node to another, possibly with significant latency. Paper [4] proposed a Predict and Relay scheme for DTN to predict the future contacts between two specified nodes at a specified time. In this approach, nodes determined the probability distribution of future contact

© Springer Nature Singapore Pte Ltd. 2017
Q. Yu (Ed.): SINC 2016, CCIS 688, pp. 126–136, 2017.
DOI: 10.1007/978-981-10-4403-8_12

times and chosen a proper next-hop in order to improve the end-to-end delivery probability. The author in [5] had proposed the design of contact plans that could minimize path costs and maximize path connecting reliability in DTN networks. In [6], the authors analyzed the problems as Contact Plan Design (CPD) for disruption-tolerant satellite networks, which were caused by resource constraints and other criteria related to capacity and fairness. As the typical solution for the routing of the spatial DTN, the Contact Graph Routing (CGR) had been proposed to contact graphs dynamically by using the contact plan and make routing decisions accordingly in [7].

Ever since CGR first appeared, the research community had worked on improving its functionality and usage. In [8], the authors provided improvements to the underlying cost function of CGR to avoid routing loops and suggest applying Dijkstra's shortest path algorithm for path selection. In [9], the authors proposed the use of source routing, and suggested that information extracted at the source node be stored in a Bundle Extension Block [10]. In [11], the authors proposed the CGR with Earliest Transmission Opportunity (CGR-ETO) scheme to improve the accuracy of predicted bundle delivery time by considering the available information on queueing delay, and also the overbooking management method to proactively handle contact oversubscription, which occurred when high priority bundles were forwarded for transmission on a contact that was already fully subscribed by lower priority bundles. In [12], the authors discussed the application of CGR in different space DTNs. For LEO satellite DTN communications, the authors in [13] investigated the suitability of CGR in two practical application scenarios: Earth observation and data mule respectively.

Due to the limited contact duration and transmission rate, contacts have a finite volume or "capacity", most of above researches handled the contact capacity of a given contact between local node and one of its neighbors depending on the priority of message. However, in many practical scenarios, traditional scheme exists many disadvantages. When the local node forwards a higher priority message to its proximate node in a contact, it deliberately neglects lower priority ones in order to enforce priority. As a result some lower priority messages which currently queue for transmission to the neighboring node will miss their contacts. Once the forfeit time of these messages exceeds in the contact end time, these messages are re-forwarded, which will lead to low throughput, high probability of message loss and some performances are not as ideal as expected. In addition, traditional scheme computes all possible routes for each message on each node and selects the best path to destination, which may result in the re-computing for some nodes and lead to a heavy routing computation burden for the network.

Aiming at the above problem, we introduce network coding into the CGR to improve system performance in satellite communications, which is few proposed among the existing CGR schemes. In recent years, network coding has been proved as an effective way to get better system performance by its advantage in improving the reliability and saving bandwidth, and its application for network routing protocols as a new paradigm. The authors in [14] proposed a distributed coding-aware routing protocol that can maximize throughput and minimize delay of real-time multimedia flows in wireless multihop networks. In [15], the authors applied network coding to multicast transport of BGP in satellite network to improve the reliability of BGP routing transport and reduce retransmission count. In [16], the authors focused on the study of security aspects of intra-flow network coding based routing system. In [17], a joint scheme was proposed by employing opportunistic network coding and opportunistic routing in wireless network, and an efficient data gathering framework for correlated source was established. In [18], the system performance was greatly impacted by the using of network coding based opportunistic routing scheme and the correlation among the links.

In light of the above discussion, we propose an improved Contact Graph Routing (CGR) scheme by introducing the Destination based Network Coding (DNC), namely DNC-CGR. In our proposed scheme, we determine whether a satellite node needs to employ network coding to cope with its message or not according to its contact residual capacity in one contact. When the contact residual capacity is enough to forward a complete message, the message will be inserted into the output transmission queue depending on the priority; when the contact residual capacity is not enough caused by that the contact is fully booked by the lower priority message, multiple messages having different destination will be inserted into the queue subscribing the contact capacity depending on message priority, and multiple messages having the same destination will be encoded into a single encoded message by network coding. To do this, the lower priority messages can be forwarded at the same time and will not miss their contacts, also the times of routing re-computed will be decreased efficiently.

The remainder of the paper is organized as follows: In Sect. 2, we have a brief introduction on the overall process of the DNC-CGR scheme. In Sect. 3, we describe the encoding process of DNC-CGR in detail. The simulation results are given in Sect. 4. Finally, conclusions are drawn in Sect. 5.

2   The Overall Process of DNC-CGR Scheme

In this section, we state the implementation process of our proposed DNC-CGR scheme.

As shown in Fig. 1, the overall process flow of DNC-CGR scheme mainly consists of four steps: forming the global contact plan, constructing the contact graph, calculating and comparing the Estimated Capacity Consumption (ECC) and the residual capacity, and forwarding messages. We assume that a global contact plan of all contacts may be

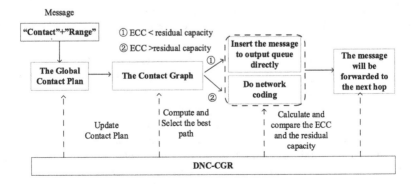

Fig. 1. The overall process flow of DNC-CGR scheme

distributed in advance to all nodes, all the contacts are planned and scheduled, rather than predicted or discovered, and each node can independently use the contact information (contact duration and capacity) to locally compute the available routes for each message.

In the following, we detailedly describe the implementation process of the DNC-CGR scheme:

Step1: Forming the Global Contact Plan. According to accurate contact plan information that are composed by contact messages and range messages, DNC-CGR provided each node the message in the form of contact plan. As time passes, the contact plan must be updated by adding the connectivity opportunities in the future, and removing the connectivity opportunities in the past.

Step2: Constructing the Contact Graph. Before routing computation, each node can construct a contact graph with a complete contact plan, and then independently use the contact graph to compute the available routes for each message and select the best path to forward its message.

Step3: Calculating and Comparing the ECC and the Residual Capacity. Before forwarding a message to the next hop in a contact, the local node will calculate and compare the ECC and the residual capacity for each contact. When the ECC is larger than the residual capacity, which is caused by that the contact is fully booked by the lower priority message, multiple messages transmitted to different destination will be inserted into the queue depending on message priority, and multiple messages having the same destination will be encoded into a single encoded message by the Random Linear Network Coding (RLNC). Otherwise, the message will be inserted into the output transmission queue depending on the priority.

Step4: Forwarding Messages. After dealing with messages in different method, the messages in the output transmission queue ordered by priority or encoded messages will be forward to next hop.

3 The Encoding Process of DNC-CGR

In this section, we have a detailed description for the encoding process of DNC-CGR before the message are forwarded to the next hop.

With the proposed DNC-CGR scheme, when the local node A is ready to forward the message P_1 to its proximate node B, DNC-CGR will compute the ECC for P_1 and the residual capacity of a given contact between node A and node B. We choose the forwarding method according to the value of the ECC and the residual capacity.

3.1 ECC < Residual Capacity

When the ECC of P_1 is less than residual capacity, and considering the particular character of inter-satellite links (ISLs) quick handover, we will insert P_1 into the output transmission queue that is ready to be transmitted to node B (depending on priority) instead of encoding.

3.2   ECC > Residual Capacity

When the ECC of P_1 is larger than residual capacity, we consider three cases as follows:

(1) There is no message having the same destination with P_1 in the current transmission queue, which is ready to be transmitted to proximate node B, the lower priority message will be discarded in this contact. As shown in Fig. 2, the priority of message Q_1 is lower than P_1, so it will miss the contact and be replaced by P_1 in the output transmission queue.

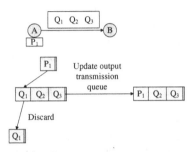

Fig. 2. Case 1: there is no message having the same destination with P_1

(2) There is only one message P_2 having the same destination with P_1 in the current queue for the transmission to proximate node B. So P_1 and P_2 will be encoded with RLNC into an encoded message as shown in (1),

$$P' = a_1 P_1 + a_2 P_2 \tag{1}$$

Where a_1, a_2 are coding coefficients which are randomly selected from finite field F_q. All the additions and multiplications are performed over the finite field. The encoded message P' including the information of P_1 and P_2 will replace P_2 to be inserted into the output transmission queue depending on priority, and then P_1 and P_2 will be discarded (Fig. 3).

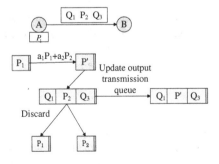

Fig. 3. Case 2: P_2 has the same destination with P_1

(3) There are two messages P_2 and P' (P' is an encoded message) having the same destination with P_1 in the current output queue that is ready to be transmitted to proximate node B. So P_1 will be encoded with P_2 and P' respectively with RLNC, and the results are given by,

$$P'' = a_3 P_1 + a_4 P_2 \qquad (2)$$

$$P''' = a_5 P_1 + a_6 P' \qquad (3)$$

We have known that each destination node maintains a coding coefficient matrix and each encoded message is a linear combination of the primitive messages (Fig. 4). At the destination, m encoded messages are be received and stored into the buffer queue. In order to decode n messages at a destination node, it is required $m \geq n$, that is to say if the destination node would like to retrieve n primitive messages, it must have received at least n linearly independent encoded messages and adopt Gaussian elimination method to solve the n linear equations. Recalling that the coefficients are randomly selected from finite field, we notice there is a high probability that all encoded messages are linearly independent, and the probability of successful decoding can reach to 99.6% as $q = 2^{16}$ [19], in a real application, it is sufficiently that $q = 2^8$ [20].

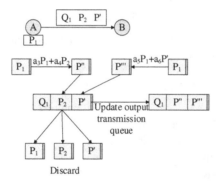

Fig. 4. Case 3: P_2 and P' have the same destination with P_1

We take an example to explain the decoding process at the destination. We assume the destination node receives linearly independent encoded messages R_1, R_2 and R_3 as shown in (4), (5) and (6),

$$R_1 = a_1 P_1 + a_2 P_2 + a_3 P_3 \qquad (4)$$

$$R_2 = a_4 P_1 + a_5 P_2 + a_6 P_3 \qquad (5)$$

$$R_3 = a_7 P_1 + a_8 P_2 + a_9 P_3 \qquad (6)$$

Then these linear equations can be conversion to a matrix form,

$$r = Ap \tag{7}$$

$$A = \begin{bmatrix} a_1 & a_2 & a_3 \\ a_4 & a_5 & a_6 \\ a_7 & a_8 & a_9 \end{bmatrix}, p = \begin{bmatrix} P_1 \\ P_2 \\ P_3 \end{bmatrix}, r = \begin{bmatrix} R_1 \\ R_2 \\ R_3 \end{bmatrix} \tag{8}$$

The primitive messages can be retrieved by using matrix inversion:

$$p = A^{-1}r \tag{9}$$

4   Simulation Results

In this section, we verify the network performance of DNC-CGR scheme, and compare it with the traditional scheme, viz. the standard CGR [7]. We mainly evaluate the network performance in a subnet of the LEO satellite DTN communication network, which has five satellite nodes, and the orbit altitude is 1400 km, orbit inclination is 48°. We set that the link transmission rate is 512 kbit/s, and consider a message payload of 100 kB in our experiments (ECC = 107235B). The detailed simulated parameters are shown in Table 1.

Table 1. Simulated parameters

Parameter name	Parameter value
Satellite altitude (km)	1400
Inclination (°)	48
Cross-seam ISLs	No
# of ISLs	2 intra + 2 inter plane
Simulation duration (s)	100
Transmission rate (Kbit/s)	512
Field size	$q = 2^8$
ECC (B)	107235

In Fig. 5, we compare the network throughput by adopting traditional scheme and DNC-CGR scheme respectively. As we can see, compared with traditional scheme, the throughput by introducing DNC-CGR scheme has a better performance and as the simulation time goes by, the throughput has a further growth. This also means that bandwidth can be used more effectively.

The Fig. 6 shows the number of messages in ISLs by introducing traditional scheme and DNC-CGR scheme respectively. The number by using DNC-CGR scheme maintains a stable state, rather than a sharply change. On the contrary, the number of messages will change quickly by using traditional scheme. The phenomenon is attributed to that the multiple messages can be encoded to an encoded message by using DNC-CGR, so the number of messages in ISLs will get a greatly reduced.

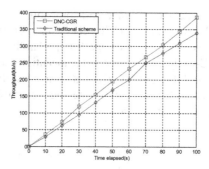

Fig. 5. Throughput of traditional scheme and DNC-CGR

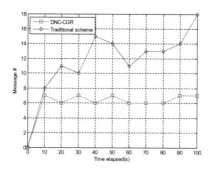

Fig. 6. Number of messages in ISLs

The average transmission times of two schemes are simulated and showed in Fig. 7. It can be easily obtained that our proposed scheme behaves much better than traditional scheme. The reason for the better performance of DNC-CGR scheme is that some lower priority messages are encoded with higher priority messages having the same destination with it, the contact residual capacity is sufficient to forward lower priority message, instead of being re-forwarded.

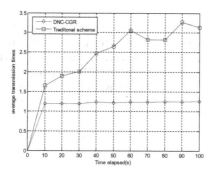

Fig. 7. Average transmission times

In Fig. 8, we investigate the message loss probability under the condition of employing traditional scheme and DNC-CGR scheme respectively. We can see that as the simulation time passes away, the message loss probability of traditional scheme ascends quickly, but DNC-CGR rises slowly. In the meantime, at each of the simulation time, the message loss probability of DNC-CGR is always less than traditional scheme. Thus it can be seen that the performance of the message loss probability of DNC-CGR scheme is better than that of traditional scheme.

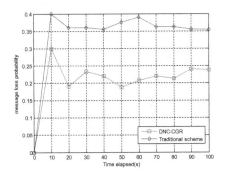

Fig. 8. Message loss probability

5   Conclusion

In this work, we have surveyed traditional routing schemes in LEO satellite DTN communications and proposed a complementary routing scheme, namely DNC-CGR, aiming at the disadvantages of the existing CGR schemes. In our proposed scheme, we determine whether a satellite node needs to employ network coding to cope with its message or not according to its contact residual capacity in one contact. Simulation results show that the proposed DNC-CGR scheme can reduce the message loss probability and the average transmission times, also increase the system throughput and improve the number of messages in ISLs in LEO satellite DTN communications.

Acknowledgment. This work was jointly supported by the National Natural Science Foundation of Major Research Project in China (No. 91438110), the National Natural Science Foundation in China (No. 61601075), the Natural Science Foundation Project of CQ CSTC (No. cstc2016jcyjA0174), the Scientific and Technological Research Program of Chongqing Municipal Education Commission (No. KJ1500440), the Natural Science Foundation Project of CQUPT (No. A2014-111).

References

1. Sacchi, C., Bhasin, K., Kadowaki, N., et al.: Technologies and applications of future satellite networking. IEEE Commun. Mag. **53**, 154–155 (2015)
2. Caini, C., Cruickshank, H., Farrell, S., et al.: Delay and disruption-tolerant networking (DTN): an alternative solution for future satellite networking applications. Proc. IEEE **99**, 1980–1997 (2011)
3. Jain, S., Fall, K., Patra, R.: Routing in a delay tolerant network. ACM Sigcomm Comput. Commun. Rev. **34**, 145–158 (2004)
4. Yuan, Q., Cardei, I., Wu, J.: Predict and relay: an efficient routing in disruption-tolerant networks. In: International Symposium on Mobile Ad Hoc Networking and Computing, pp. 95–104. ACM, New York (2009)
5. Fraire, J.A., Madoery, P.G., Finochietto, J.M.: On the design and analysis of fair contact plans in predictable delay-tolerant networks. IEEE Sens. J. **14**, 3874–3882 (2014)
6. Fraire, J.A., Finochietto, J.M.: Design challenges in contact plans for disruption-tolerant satellite networks. IEEE Commun. Mag. **53**, 163–169 (2015)
7. Burleigh S.: Contact graph routing (2010). http://tools.ietf.org/html/draft-burleigh-dtnrg-cgr-01
8. Segui, J., Jennings, E., Burleigh, S.: Enhancing contact graph routing for delay tolerant space networking. In: IEEE Global Telecommunications Conference (GLOBECOM), pp. 1–6. IEEE Press, Houston (2011)
9. Birrane, E., Burleigh, S., Kasch, N.: Analysis of the contact graph routing algorithm: bounding interplanetary paths. Acta Astronaut. **75**, 108–119 (2012)
10. Scott, K., Burleigh, S.: Bundle protocol specification (2007). http://tools.ietf.org/html/rfc5050
11. Bezirgiannidis, N., Caini, C., Padalino, D.D.M., et al.: Contact graph routing enhancements for delay tolerant space communications. In: Advanced Satellite Multimedia Systems Conference and the Signal Processing for Space Communications Workshop (ASMS/SPSC), pp. 17–23. IEEE Press, Livorno (2014)
12. Araniti, G., Bezirgiannidis, N., Birrane, E., et al.: Contact graph routing in DTN space networks: overview, enhancements and performance. IEEE Commun. Mag. **53**, 38–46 (2015)
13. Caini, C., Firrincieli, R.: Application of contact graph routing to LEO satellite DTN communications. In: IEEE International Conference on Communications (ICC), pp. 3301–3305. IEEE Press, Ottawa (2012)
14. Jeong, M., Ahn, S., Oh, H.: A network coding aware routing with considering traffic load balancing for the multi-hop wireless networks. In: IEEE International Conference on Information NETWORKING (ICOIN), pp. 382–384. IEEE Press, Kota Kinabalu (2016)
15. Han, W., Wang, B., Feng, Z., et al.: NCSR: multicast transport of BGP for geostationary satellite network based on network coding. In: IEEE Aerospace Conference, pp. 1–10. IEEE Press, Big Sky (2015)
16. Cai, R.J., Ali, G.G.M.N., Aung, C.Y., et al.: simulation study of routing attacks under network coding environment. In: IEEE International Conference on Communications (ICC), pp. 1–6. IEEE Press, Kuala Lumpur (2016)
17. Tan, C., Zou, J., Wang, M.: Joint opportunistic network coding and opportunistic routing for correlated data gathering in wireless sensor network. In: IEEE Conference on Vehicular Technology (VTC-Fall), pp. 1–5. IEEE Press, Las Vegas (2013)

18. Khreishah, A., Khalil, I., Wu, J.: Universal network coding-based opportunistic routing for unicast. IEEE Trans. Parallel Distrib. Syst. **26**, 1765–1774 (2015)
19. Ho, T., Medard, M., Shi, J., et al.: On randomized network coding. In: Proceedings of the 41st Allerton Annual Conference on Communication, Control, and Computing, pp. 1–3. IEEE Press, Monticello (2003)
20. Gkantsidis, C., Miller, J., Rodriguez, P.: Anatomy of a P2P content distribution system with network coding. In: Proceedings of the 5th International Workshop on Peer-to-Peer Systems, pp. 1–6. IEEE Press, Santa Barbara (2006)

A Trust Holding Based Secure Seamless Handover in Space Information Network

Zhuo Yi[1(✉)], Xuehui Du[1], Ying Liao[2], and Lifeng Cao[1]

[1] The Forth School, Information Engineering University, Zhengzhou, China
yizhuo_513@163.com
[2] Military Operation Research, National Defence University, Beijing, China

Abstract. Reducing handover delay at the same time of maintaining high level of handover security is always a challenging issue, especially for secure handover in space information network where resources are limited and links are highly exposed and intermittent connected. To address this issue, this paper focuses on reducing handover authentication delay and adopts the secure context transfer method as a replace for handover authentication. To achieve this purpose, a trust holding based secure seamless handover scheme is proposed. In this scheme, a full trust holding mechanism based on finite state machine is firstly proposed to build a light-weighted trust system. Then a secure trust state context transfer method is given based on the full trust holding mechanism. Instead of handover authentication, if a mobile node is trusted by current AP, history trust state information is sent to target AP to verify MN's identity. A pre-authentication method is introduced as a complement to cope with the situation where MN is not trusted by current AP. With these, a node can easily build a trust relationship with target APs while handover begins. Security proofs and performance analysis demonstrate the effectiveness and efficiency.

Keywords: Seamless handover · Secure handover · Trust holding · Near space · Secure context transfer

1 Introduction

Currently, Space Information Network (SIN) has become the focus of national and international research due to its important strategic position and unique advantages of various information accessing and multiple task collaboration. Different from wired network in the ground, nodes in SIN suffer from frequent handover due to the high-speed moving of SIN mobile nodes (MN), inevitably leading to high delays.

Meanwhile, due to the SIN's natures of highly exposure, strong signal interference, intermittent links and dynamic changing MN locations, communication information in SIN could be easily captured by attackers, and MN or Access Point (short for AP,

This work is supported by the National Natural Science Foundation of China (No. 61502531, No. 61403400).

© Springer Nature Singapore Pte Ltd. 2017
Q. Yu (Ed.): SINC 2016, CCIS 688, pp. 137–150, 2017.
DOI: 10.1007/978-981-10-4403-8_13

including LEO/MEO satellites or near aircrafts) could be forged to reach certain purposes. Especially during a handover procedure when MN and new AP do not know each other, a fake MN can easily access the New AP and further collapse the whole SIN, or a forge AP could redirect a MN to nonexistent AP. Both of them may further cause severe secure threats during handover such as DoS, Man-in-the-middle attacks and forgery attacks, etc.

Therefore, security authentications for handover are urgently required at the same time of reducing handover delays. However, authentication is a time-consuming operation and a compromise has to be made between efficiency and security. Among existing secure handover solutions addressed, the main idea is combining authentication mechanism with efficient seamless based on movement destination prediction, light-weighted authentication algorithms and security context transfer method.

Provided that nodes in SIN have more limited storage capacity, unstable links and higher speed, these solutions can't be directly applied to SIN and changes have to be made for SIN. This paper introduces a full trust holding mechanism to build a light-weighted trust system and proposes a trust holding secure seamless handover (TH-SSH) scheme based on the ideas of secure context transfer mechanism and a pre-authentication.

Works and contributions of this paper include:

(1) A full trust holding mechanism based on finite trust state machine is proposed to guarantee MNs' trust throughout its whole lifetime. Once a potential threat that may lead to trust attenuation occurs, an authentication based on message piggyback is triggered to refresh trust state.
(2) A secure trust state context transfer (STSC) method is given as a replace of handover authentication. This method transfers MNs' trust value to new APs to establish new trust relationship, which reduce much handover delay in the same time of ensure security.
(3) A TH-SSH scheme combining STSC meth-od and pre-authentication method is proposed. STSC method and pre-authentication method are complementary in building trust relationship be-tween MNs and APs.

The rest of this paper is organized as follows. Section 2 describes the related works and the specific requirement of secure handover in SIN. The secure handover models in SIN which includes network topology and secure handover model are covered in Sect. 3. In Sect. 4, the TH-SSH scheme is addressed followed by security proofs and performance evaluation in Sect. 5. Finally, conclusions are drawn in Sect. 6.

2   Related Works

Secure Seamless handover is a communication switch process when a MN changes its access point, during which services with no user perceivable interruptions and security level are both guaranteed.

To provide secure seamless handover, existing literature have almost focus on reducing authentication delays and optimizing handover process. From the perspective of efficiency of authentication, light-weighted authentication algorithms [1, 2] and authentication protocols [3] with zero communications have been proposed. For the idea of optimizing handover process to decrease authentication delays, pre-authentication [4] and proactive handover mechanisms [5] have been addressed. In addition, Security Context Transfer (SCT) mechanism is proposed for a replacement of authentication process. Security Context Transfer (SCT) mechanism has also been studied in [6–9]. These SCT schemes have the advantage of needing no communication between an AAA server and an AP. Furthermore, batch signature verification scheme [10] is proposed for APs to reduce overall handover efficiency by simultaneously verify multiple users' identities.

To provide seamless handover while maintaining the security level, it is preferable to transfer security context information from one network to other in a timely fashion [6]. Security context contains state of authentication and authorization, cryptographic keys etc. It is used to support trust relation of and to provide communication security for network nodes/entities. The context is typically shared between a pair of network nodes or networks [7].

3   Problem Description of Secure Handover in SIN

3.1   Network Topology of SIN

Take handovers in Near Space for simplification, the handover model can be simplified as Fig. 1. In the network topology of Near Space Net-work, shown in Fig. 1, two layers of APs are included: LEO core network denoting LEO layer and NSN1 or NSN2 repre-senting near space layer. The former consists of several homogeneous or heterogeneous LEO satellites serving as routers of nodes in near space layer, and the latter consists of heterogeneous aircrafts such as aerostats, airship, floating air-ball, which is represented by STA1 to STA 4. Heterogeneous aircrafts usually play the role of gateway for MN in near space. It can be assumed that STA1 and STA2 belong to domain NSN1, while STA3 and STA 4 are owned by domain NSN2. Meanwhile, STA2 and STA3 are assumed to be heterogeneous APs due to diversity of near space aircrafts.

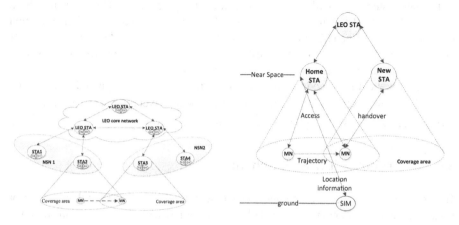

Fig. 1. Topology of SIN   **Fig. 2.** Handover model in SIN.

3.2   Handover Model in SIN

According to above network topology and a normal handover process, entities in our handover model (Fig. 2) involved in a handover include home station (Home STA), new station (New STA), mobile node (MN), space station in LEO layer and Satellite Information Manager (SIM). Each type of nodes has its own properties and is responsible for certain tasks.

MN is the subject of handover which could be unmanned aircraft, ship, plane or other mobile objects that have to access SIN to get informed of emergent message or on-line assistance. It's assumed that MN do not make handover decision but just request for handover once entering signal overlap zone of different APs. What's more, MNs' flying meets certain mobile models with which the future location could be easily predicted.

SIM, which is deployed on the ground for managing all LEO satellites' status information, takes charge of nodes' location prediction as well. Thus, future location of MN could be sent to Home STA for handover decision making.

The benefits of such design are that handover decision could be made ahead of time and authentication based on secure context transfer or pre-authentication could be implemented in advance so as to reduce handover delays. Additionally, such a design is justifiable when ground Aids is supported in the future SIN, especially when certain special missions have to be performed. Meanwhile, STA (Home STA and New STA) are approximately stationary relative to the Earth, and have a relatively stable link to SIM which provides a better data transferring QoS.

Home STA and New STA, APs of SIN, are separately the source and the destination of handover. There are STAs in two layers in this handover model, near space STAs and LEO STA. All these STAs serve as service providers that are responsible for handover decision to choose best AP, handover authentication to verify node identities and storing MN's access state information and relay nodes that transfer communication messages between MNs and New STAs. Moreover, different STAs share a private key group to build trust relationship.

According to above descriptions, a handover procedure could be briefly addressed as following steps: After MN passes the access authentication and accesses into SIN, SIM and Home STA together make handover decision based on MN location prediction provided by SIM. When MN moves to signal overlap zone of redundant APs, a handover event is triggered and a handover request is sent to Home STA. Then MN and New STA try to execute handover procedure though Home STA.

To ensure the trust of nodes and secure data transmission, it's assumed that MN and Home STA share a session key (key_{mh}) STAs of the same layer share a private key group $\{Ks_1, Ks_2, \ldots, Ks_n\}$, SIM and STAs share a session key (Key_{ss}).

4   The Proposed TH-SSH Scheme

Generally, conventional secure handover methods are based on pre-authentication or proactive handover. However, these methods are inefficient enough to perform handover procedures due to the high-speed of SIN nodes and the time-consumption of authentication process. This leads to an increase in handover delays and handover failures, resulting in poor QoS or even availability of SIN. The proposed TH-SSH scheme promises secure seamless handover with shorter handover latency and stronger security trust mechanism. It introduces a light-weighted trust holding mechanism to guarantee that a MN is trusted throughout the whole lifetime accessing to SIN. Meanwhile, a method combining security context transfer and pre-authentication is adopted to build trusted relationship between MNs and New STA, which largely reduces mutual authentication time. Therefore, the TH-SSH scheme consists of two main parts: a full trust holding mechanism and Trust holding based secure seamless handover scheme.

4.1 Full Trust Holding Mechanism

Definition 1 Trust Value (TV) denotes to a double value between 0 and 1 that depicts a node's trust degree. Trust State (TS) is a binary variable that represents whether a node is trusted. Given a threshold θ, if TV is great to θ, TS equals to 1 and denotes a trusted state, otherwise TS is 0 and an untrusted state.

$$TS = \begin{cases} 1, & TV \geq \theta \\ 0, & TV < \theta \end{cases} \tag{1}$$

Full trust holding means that keep the TS of a MN to be 1 during the time interval between the time it accesses the SIN and the time it quits the SIN.

Typically, nodes in SIN suffer from intermittently connected links, highly exposed network environments and dynamic organized SIN, which give attackers many opportunities to invade into SIN and leave great challenges to nodes management. So as to prevent SIN from potential attacks, the identities of nodes should be trusted all throughout their lifetime. Simply put, nodes in SIN should be verified once they become untrusted during their lifetime.

With deep insight into the whole lifetime, there are five cases where secure threats may occur. (1) When a MN first accesses SIN, its TS may be untrusted. (2) During the interval that links between nodes disconnect, a fake node may disguise as a legal node. (3) A forged node may send a quit request to AP to isolate a MN or even launch a DoS. (4) At the time of handover, a malicious MN may try to access SIN or a forged AP may pretend to be legal to redirect a victim MN to dangerous situations. Moreover, (5) nodes with ulterior motives may monitor SIN links to capture private keys. All above cases may cause a MN to be untrusted.

Therefore, a full trust holding mechanism based on proper authentications is proposed and several kinds of authentication steps have been adopted to judge whether a MN could be trusted, which include access authentication (AA), refresh authentication (RA), handover authentication (HA), and quit authentication (QA), as is shown in Fig. 3. All these authentications are implemented based on message piggyback.

Before a MN accesses SIN, the TV is initialized to 0 to mark its untrusted. When the MN passed these authentications, the MN is regarded as trusted and the TV is set to 1. Otherwise, if the MN failed, an accelerated decrease in TV occurs as a penalty. Each MN has three successive failures at most before being isolated.

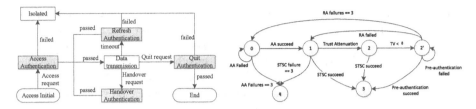

Fig. 3. A message piggyback based full trust holding mechanism

Fig. 4. Trust state transfer graph or trust state machine

From those five cases and the full trust holding mechanism, it could be drawn that passage of time, authentication failures and link disconnections are the key factors leading to mistrust. Though transmission encryption is a useful measure to guarantee data integrity, it brings little impact on TV. So transmission encryption is not taken into account when calculating TV. Accordingly, a trust attenuation model based on these factors could be given in Eq. 2.

$$TV_t = TV_0 \times (1 - \gamma_1 N_1) \times (1 - \gamma_2) \cdot T_0 - N_1/3 \tag{2}$$

Where TV_0 is the TV since last authentication of a MN, γ_1 denotes the attenuation coefficient of disconnection, N_1 represents the times of links disconnection between APs and MNs, γ_2 refers to as the attenuation coefficient of time, and N_1 is the number of authentication failures.

According to the full trust holding mechanism and the trust attenuation model, the TV of a MN varies and its TS could be described by a trust state machine (TSM) in Fig. 4. This TSM could be depicted by a quintuple $(\sum, S, S_0, \delta, F)$.

$$TSM = \left(\sum S, S_0, \delta, F \right) \tag{3}$$

Where $\sum$ denotes inputs set which includes return of authentication. Taking access authentication for instance, if a MN passes AA, then the input is AA_1, otherwise AA_0. Thus 4 pairs of inputs are added into $\sum$ as Eq. 4 shows. Moreover, trust attenuation could also lead to TS transition. Thus an input TT indicating trust attenuation is included.

$$\sum = (AA_0, AA_1, RA_0, RA_1, HA_0, HA_1, QA_0, QA_1, TT) \tag{4}$$

And S refers to a TS set described in Eq. 5, where S_0 represents the initial TS of MN S_0, and S_q describes the end TS F. In this TSM, trusted state is divided into two TSs (Access Trusted State depicted by S_1 and Handover Trusted State S_3) given the difference between access and handover. And a trust attenuation state S_2 is added to S as a transitional state between S_1 and mistrusted state $S_{2'}$.

$$S = (S_0, S_1, S_2, S_{2'}, S_3, S_q) \tag{5}$$

Meanwhile, δ indicates a TS transition function as Eq. 6 describes. Once given an input $\varepsilon|\varepsilon \in \Sigma$, TS of the MN changes from S_{n-1} to S_n.

$$\delta : S_n = f\left(S_{n-1}, \varepsilon\right) \tag{6}$$

Based on the full trust holding mechanism and the TSM, a full trust holding algorithm could be given.

Algorithm 1.   The full trust holding algorithm

```
1: set TS_mn = S_0; call AA;
2: if Num(AA)>3, set TS_mn = S_q and move to 7
   if ret(AA) == 1, set TS_mn = S_1; call attenuation;
   if ret(AA) == 0, set TS_mn = S_0; call AA;
3: Calculate TV = TV_0 × (1 − γ_1 N_1) × (1 − γ_2) · T_0;
   if TV· set TS_mn = S_2 moves to 3;
   if TV≤ set TS_mn = S_2, call RA, moves to 4;
4: if Num(RA)>3, set TS_mn = S_q and move to 7
   if ret(RA) == 1, set TS_mn = S_1; call attenuation;
   if ret(AA) == 0, set TS_mn = S_0; call RA;
5: If handover event occurs, Call HA
6: if Num(HA)>3, set TS_mn = S_q and move to 7
   if ret(HA) == 1, set TS_mn = S_1; call attenuation;
   if ret(HA) == 0, set TS_mn = S_0; call HA;
7: set TS_mn = S_0; end
```

4.2   Trust Holding Based Secure Seamless Handover Scheme

With the full trust holding mechanism deployed on APs, the TV of MNs could easily be stored in APs. Trust relationship could be established between APs and MNs by transferring TV from Home STA to New STA. The main idea of TH-SSH mechanism if that when a MN requests for handover, Home STA reads TV of the MN and judges whether the TV is greater than the trusted threshold θ. If greater, a secure trust state context transfer is adopted to instead of handover authentication. Otherwise, a pre-authentication is applied to verify the legality of MN's identity.

4.2.1   Secure Trust States Context Transfer Method

[6] proves that secure context transfer is more preferable than pre-authentication even in inter-domain handovers possessing unanimous authentication system. However, ticket-based or random nonce-based context cannot meet the trust level because the very malicious node may monitor the communication links and resend the context to obtain New AP's trust. The secure context in STSCT combines the TV of MN and challenge value (CV) to inform New STA with the history TS of MN before handover.

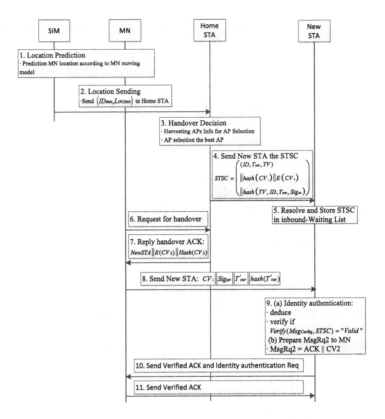

Fig. 5. Secure trust state context transfer method

The secure context in STSCT could be de-fined by the vector STSC as follows:

$$STSC = \begin{pmatrix} ID_{MN}, TV_{MN}, T_{cur}, H(CV_1), E(CV_2), \\ H(TV_{MN}, T_{cur}, ID_{MN}, Sig_{MN}) \end{pmatrix} \qquad (7)$$

Where TV_{MN} is TV of the MN, and T_{cur} denotes current timestamp, ID_{MN} indicates the identity of the MN. Meanwhile CV_1 is a challenge value for New STA to verify the MN, and CV_2 is the challenge value for MN to verify the New STA, Sig_{MN} is the signature of the MN. Furthermore, $(TV_{MN}, T_{cur}, ID_{MN}, Sig_{MN})$ and CV_1 are encrypted by hash function to prevent them from being forged. CV_1 is encrypted by Key_{HN} a private key shared by Home STA and New STA.

As shown in Fig. 4, STSCT method is triggered when a MN requests for handover at S_1 and S_2 where the MN is trusted. Figure 5 illustrates the whole process of STSCT method.

Step1. The method starts from MN location prediction based on mobile model. This location prediction is made by MN, SIM or both of them cooperatively.

Step2. SIM generates $Loc_{MN} = GeoHash(x, y)$ using GeoHash algorithm [11] for compressing traditional location description and convenient APs searching during AP selection phase. Then SIM sends location Msg_{Loc} to Home STA encrypted by Key_{SH}. SIM → Home STA:

$$Msg_{Loc} = \{Loc_{MN}||ID_{MN}\}_{Key_{ss}}$$

Step3. Home STA decrypts the Msg_{Loc} using Key_{SH} and harvests APs information based on Loc_{MN}. Then it makes AP selection decision according to APs information and access requirements. MN's access requirements are known to Home STA for that Home STA has been its AP since last handover.

Step4. With the TV of MN, Home STA chooses an authentication method and if the STSCT method is selected, Home STA sends New STA the secure trust state context Msg_{stsc} for MN identity verification.

$$Msg_{stsc} = ID_{MN}||TV_{MN}||T_{cur}||H(CV_1)||E(CV_2)||H(TV_{MN}, T_{cur}, ID_{MN}, Sig_{MN})$$

Home STA → New STA: $\{Msg_{stsc}\}_{Ks_i}$

Step5. New STA receives Msg_{stsc} and resolves it to get each components of the upcoming MN. Then New STA stores the information of $(ID_{MN}, TV_{MN}, T_{cur})$, $H(TV_{MN}, T_{cur}, ID_{MN}, Sig_{MN})$ and $H(CV_1)$ in unbound-HO-waiting list for future MN authentication. And it also decrypts $E(CV_2)$ to get CV_2 for proving its identity to the MN.

Step6. While the MN moves into the overlap area of redundant APs, a handover request is triggered and sent to Home STA.

MN → Home STA: $\{Msg_{req}\}_{key_{mh}}$

Step7. Home STA receives 〖Msg〗_req and replies to MN the request ACK, then sends New STA's information including access parameters and challenge value.

$$Msg_{rpt} = Param_{NSTA}||H(CV_2)||E(CV_1, Key_{mh})||ACK$$

Home STA → MN: $\{Msg_{rpt}\}_{key_{mh}}$

Where CV_2 is adopted for verifying the New STA and $Param_{NSTA}$ describes the access parameters.

Step8. MN gets Msg_{rpt} and resolves $Param_{NSTA}$ and CV_2 corresponding to New STA. And then MN creates a message Msg_{auth} with its own signature, challenge value and hash digest of current time.

$$Msg_{auth} = ID_{MN}||Sig_{MN}||CV_1||T'_{cur}||H(T'_{cur})$$

MN → New STA: $\{Msg_{rqt}, Msg_{auth}\}$

Step9. From the message received, New STA searches its inbound-HO-waiting list for MN's handover information and executes following steps to verify the trust state of MN:

(a) New STA computes the hash digest of T'_{cur} and then compares the consequence hash(T'_{cur}) to $H(T'_{cur})$. If hash(T'_{cur}) equals to $H(T'_{cur})$, move to step (b).

(b) Compute the hash digest of CV_1 to get hash(CV_1), and verify whether hash(CV_1) = $H(CV_1)$ is established. If yes, move to step (c)

(c) Compute the hash digest of $(ID_{MN}||TV_{MN}||T_{cur})$ received from Home STA and Sig_{MN} from MN to get hash($ID_{MN}, TV_{MN}, T_{cur}, Sig_{MN}$). Then compare hash($ID_{MN}, TV_{MN}, T_{cur}, Sig_{MN}$) to $H(TV_{MN}, T_{cur}, ID_{MN}, Sig_{MN})$, if equal, the authentication is passed and MN is trusted by New STA.

Step10. After verifying MN's TS, New STA sends an authentication ACK and provides its own identity proof $CV_2||ID_{New}$.

New STA → MN: $Msg_{authNew} = (ACK \parallel CV_2 \parallel ID_{New} \parallel T''_{cur} \parallel H(T''_{cur}))$

Step11. MN gets ACK and CV_2, and then check whether hash(CV_2) = $H(CV_2)$. If yes, New STA passes the authentication and MN and New STA could begin the following phases. Then handover authentication phase ends.

In the above procedure, the first five steps belong to handover preparations and they are completed before the MN's handover request. Step 6 to step 11 is handover authentication phase and these steps are based on hash digest comparison which cost little time. This STSCT method has made great improvement in reducing handover delays.

4.2.2 Pre-authentication Mechanism

While the MN is not trusted by Home STA, such as S_2 shown in Fig. 4, pre-authentication mechanism is triggered when handover request occurs. The same with STSCT method, Pre-authentication mechanism also is based on location prediction in handover preparation phase. The differences begin from step 4 and steps for pre-authentication method are presented in Fig. 6.

Step1–step3 are the same to that of STSCT method and no repeat here.

Step4. With the TV of MN, Home STA chooses Pre-authentication mechanism for authentication for the TV of MN is lower to threshold θ. Home STA generates a short-term key Key_{st} based on MN's context materials and sends New STA Key_{st} for authentication in a short term.

Home STA → New STA: $Msg_{stsc} = ID_{MN}||Key_{st}||H(Key_{st})||T_{cur}||H(T_{cur})$

Step5. While the MN moves into the overlap area of redundant APs, a handover request is triggered and sent to Home STA.

MN → Home STA: Msg_{req}

Step6. Home STA receives Msg_{req} and replies to MN the request ACK, a short-term key Key_{st} and New STA's information including access parameters and challenge value.

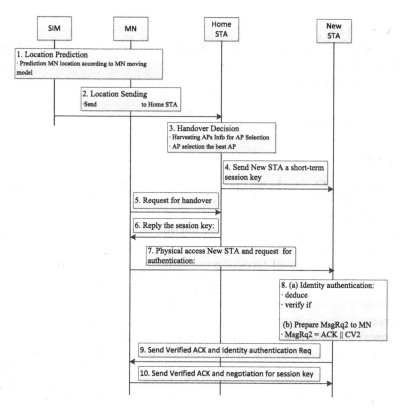

Fig. 6. Pre-authentication method

Home STA → MN: $Msg_{rpt} = ACK||Param_{NSTA}||Key_{st}||H(Key_{st})$

Step7. MN gets Msg_{rpt} and resolves Key_{st} and $Param_{NSTA}$. And then MN first physically accesses to New STA according to $Param_{NSTA}$. Then it generates a signature with its ID and current time. And it creates a signed message Msg_{sig} and sends authentication message to New STA.

$$Msg_{auth} = Rqt||ID_{MN}||T_{cur}||Sig_{MN}(ID_{MN}, T'_{cur})$$

MN → New STA: $\{Msg_{auth}\}_{Key_{st}}$

Step8. From the message received, New STA verifies the signature. And if the signature of MN is verified, New STA generates its own signature with its ID and current time.

Step9. Then a mutual authentication request message Msg_{mar} consisting of authentication ACK, its own authentication request and signature is sent to MN.

$$Msg_{mar} = Rqt||ID_{MN}||T_{cur}||Sig_{MN}(ID_{MN}, T''_{cur})$$

New STA → MN: $\{Msg_{mar}\}_{Key_{st}}$

Step10. After verifying the signature from New STA, MN sends an authentication ACK to New STA for negotiating new session key.

New STA → MN: $Msg_{authNew} = (ACK)_{Key_{st}}$

5   Experimental Evaluation

5.1   Security Proofs

According to the TH-SSH scheme, trust relationship between MN and New STA could be easily proved by the following step. To prove the security of handover, a challenge model should be given first.

First, before MN requests for handover, full trust holding mechanism is applied to make sure that MN is trusted throughout its lifetime in SIN.

Additionally, the proposed TH-SSH possesses the following security features:

(1) Guaranteeing full trust. The trust value is monitored by Home STA. Once the trust value falls below the trust threshold θ, a refresh authentication is triggered to for identity verification. Only when the MN passes the authentication, the trust value can be reset to 1. Otherwise, if authentication fails, the trust value decreases swiftly. If three authentications are tried, the TV is set to 0 and the MN is isolated from SIN. In addition, AA, QA and HA are used to prevent MNs from pretending to be trust nodes.

(2) Supporting mutual authentication. Both secure trust state context transfer mechanism and pre-authentication mechanism in TH-SSH scheme provide mutual authentication. In secure trust state context transfer mechanism, Home STA sends New STA a hash value of challenge value CV_1 and a vector of $TV_{MN}, T_{cur}, ID_{MN}, Sig_{MN}$ and its hash digest. The MN has to provide correct CV_1, T_{cur} and Sig_{MN} to prove itself. These values together verify the MN's identity. Similarly, Home STA sends MN a hash value of challenge value CV_2 which is used by New STA to prove itself. Thus, a mutual authentication is completed.

(3) Anti-replay attack. The timeliness of messages between MNs and STAs such secure trust state context, authentication massages and handover request are guaranteed by random nonce and its hash digest. For instance, step 4, step 8 and step 10 in secure trust state context transfer mechanism and step 4, step 8 and step 10 in pre-authentication mechanism have adopted random nonce to prevent from replay attacks.

(4) Defending Man-in-the-middle attack. All data transferred between MN and Home STA of between Home STA and New STA are encrypted by their private keys (Such as Key_{SH} and Key_{st}) or hash algorithms. Man-in-the-middle could not capture sensitive information.

(5) Anti-tampering attacks. The integrity of messages between MNs and STAs are guaranteed by its hash digest. Any alteration of the message will be found to prevent message tampering attacks.

5.2   Performance Analysis

Simulations based on OPNET tools are carried out to check the availability and effi-ciency of the Proposed TH-SSH scheme. Authentication Latency and Handover Latency are the focus of our attention. Simulation parameters are illustrated in Table 1. And LWA based on light-weighted authentication in [3], FA-IDH based on secure context transfer in [6] and TH-SSH are selected for comparative experiment.

Table 1. Main simulation parameters

Parameters	Values
Number of aerostats/aircrafts ·	50/300
Height of aerostats/aircrafts (km)	35/25
Coverage radius of aerostats (km)	40
Flight speed of aircrafts (km/s)	0.4
Maximum residence time (min)	3.33
Call duration (s)	180
Call arrival rate (10^{-4}calls/s)	3.5
Simulation time (s)	9000

Firstly, the Authentication Latency is tested. Authentication latency is the time interval between an authentication request and the completion of access authentication. It could be seen in Fig. 7 that the authentication delay of TH-SSH is lower than the other two schemes. There exist two reasons. The first is that compared to light-weighted authentication, secure context transfer costs less computing time and experiences less message exchange. The second reason is that secure trust state context is sent to target AP before handover request, while secure context in FA-IDH is transferred after hand-over request. Thus, TH-SSH is better in reducing authentication delays.

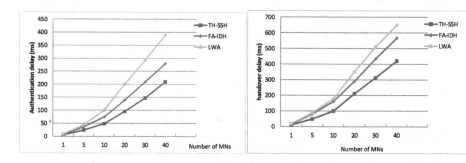

Fig. 7. Authentication latency   **Fig. 8.** Handover latency

Secondly, the Handover Latency is checked. Handover latency is the time interval between completion of link layer handover completion and the time MN receives data from other nodes again. As is shown in Fig. 8, the Handover delay of TH-SSH is lower than the other two schemes. Given the MN location prediction, execution time of phases in TH-SSH may be overlapped, which mean some phases are parallel. Meanwhile, difference in authentication time increases the advantage in handover efficiency.

6   Conclusion and Future Work

The paper is focused on secure handover in space information network and introduces a full trust holding mechanism to build a light-weighted trust system, with the support of which a trust holding secure seamless handover (TH-SSH) scheme based on the ideas of secure context transfer mechanism and a pre-authentication is put forward. A secure trust state context transfer method based on traditional secure context transfer is given as a replace of handover authentication, which would reduce much handover delays in the same time of ensure security. And a pre-authentication is adopted as a complement.

Given that the full trust holding mechanism may cause extra computing resources and time though it's performed based on message piggyback, our main focus in the future is on authentication algorithm optimization to reduce the authentication delay so as to release the burden of access points and reduce computing cost when executing trust holding mechanism.

References

1. Zhang, Z., Boukerche, A., Ramadan, H.: Design of a lightweight authentication scheme for IEEE 802.11p vehicular networks. Ad Hoc Netw. **10**(2), 243–252 (2012)
2. Cheikhrouhou, O., Koubaa, A., Boujelben, M., et al.: A lightweight user authentication scheme for Wireless Sensor Networks. In: IEEE/ACS International Conference on Computer Systems and Applications, pp. 1–7. IEEE (2010)
3. Huang, C.H., Wu, H.H., Huang, Y.J., et al.: Lightweight authentication scheme for wireless sensor networks. In: Proceedings of Global High Tech Congress on Electronics, pp. 70–74 (2012)
4. Lopez, R., Schulzrinne, H.: A framework of media-independent pre-authentication (MPA) for inter-domain handover optimization (2011)
5. Zhang, Y., Chen, X., Li, J., et al.: Generic construction for secure and efficient handoff authentication schemes in EAP-based wireless networks. Comput. Netw. **75**(10), 192–211 (2014)
6. Wang, H., Prasad, A.R.: Fast authentication for inter-domain handover. In: Proceedings of Telecommunications and NETWORKING - ICT 2004, International Conference on Telecommunications, Fortaleza, Brazil, 1–6 August 2004, pp. 973–982 (2004)
7. Wang, H., Prasad, A.R.: Security context transfer in vertical handover. In: IEEE on Personal, Indoor & Mobile Radio Communications, PIMRC 2003, vol. 3, pp. 2775–2779 (2003)
8. Georgiades, M., Akhtar, N., Politis, C., et al.: AAA context transfer for seamless and secure multimedia services over all-IP infrastructures (2004)
9. Mukherjee, R.P., Sammour, M., Wang, P.S., et al.: Methods and apparatus to facilitate data and security context transfer, and re-initialization during mobile device handover: US. US 20080240439 A1 (2008)
10. Zhou, Y., Zhu, X., Fang, Y.: MABS: multicast authentication based on batch signature. IEEE Trans. Mob. Comput. **9**(7), 982–993 (2010)
11. Malensek, M., Pallickara, S., Pallickara, S.: Fast, Ad Hoc query evaluations over multidimensional geospatial datasets. IEEE Trans. Cloud Comput. **5**(1), 28–42 (2015)

Modulation Index Selection Strategy for Quasi-Constant Envelope OFDM Satellite System

Cheng Wang[1,2,3(✉)], Yizhou He[4], Gaofeng Cui[1,2], and Weidong Wang[1,2]

[1] Key Laboratory of Universal Wireless Communications, Ministry of Education, Beijing, China
{wangcheng,cuigaofeng,wangweidong}@bupt.edu.cn
[2] Information and Electronics Technology Lab, School of Electronic Engineering,
Beijing University of Posts and Telecommunications, Beijing, China
[3] The Science and Technology on Information Transmission and Dissemination
in Communication Networks Laboratory, Shijiazhuang, China
[4] China Academy of Information and Communications Technology, Beijing, China

Abstract. Satellite communication is an important part of communication networks. Since the energy and the spectral of satellite system are limited, the energy efficiency and the spectral efficiency become two key performance metrics. However, these two metrics always increase in an opposite way. Quasi-constant envelope OFDM technology can provide high data rate communication links and maintain the peak-to-average power ratio (PAPR) less than 3 dB. It provides a way to get balance between these two metrics and it is suitable for satellite communication system. As the phase modulation method is used in quasi-constant envelope OFDM technology, the energy efficiency and the spectral efficiency of the system are all related to the modulation index. Since a typical satellite communication system contains various types of services, the problem is how to choose the appropriate modulation index according to the demands and conditions of different users. In this paper, a system performance model is proposed and it considers the energy efficiency and the spectral efficiency as two factors. The modulation index selection strategy is to choose the appropriate modulation index to maximize the system performance. The relationship between these two factors and the modulation index are analyzed and the system performance curves are given through simulation.

Keywords: Quasi-constant envelope OFDM · Satellite communication · Energy efficiency · Spectral efficiency · Modulation index selection

1 Introduction

Satellite communication has been widely used in many fields, such as seamless access and emergency communication. Various satellite communication networks have been already established. For example, the typical civilian satellite mobile communication systems contain the INMARSAT, the Thuraya and the Globalstar. These systems can support variety types of communication services. Since the energy of the satellite is collected by solar panels, the energy efficiency of satellite system is a important performance metric. Besides, the transmission rate demands of users is increasing. Since the

© Springer Nature Singapore Pte Ltd. 2017
Q. Yu (Ed.): SINC 2016, CCIS 688, pp. 151–161, 2017.
DOI: 10.1007/978-981-10-4403-8_14

available bandwidth is limited, the spectral efficiency of satellite system is another important performance metric.

As the power amplifier (PA) used in satellite communication system is always nonlinear, it requires a low peak-to-average power ratio (PAPR) transmitted signal to avoid the using of large input power backoff (IBO) and to achieve a high energy efficiency. Several constant envelope or quasi-constant envelope (i.e., signals with a PAPR less than 3 dB) transmission technologies have been proposed, e.g. the Gaussian Minimum Shift Keying (GMSK) [1], the Feher Quadrature Phase Shift Keying (FQPSK) [2] and the Constant Envelope OFDM (CEOFDM) [3]. However, the spectral efficiency of these technologies are limited and these technologies are difficult to meet the growing high speed demands of users in satellite communications.

Orthogonal frequency-division multiplexing (OFDM) technology achieves a high spectral efficiency by transmitting information through subcarriers [4]. However, its main drawback is the high PAPR [5]. A large IBO will seriously reduce the energy efficiency. Although various PAPR reduction schemes have been proposed for OFDM [6], these schemes either decrease the spectral efficiency or increase the complexity. Therefore, the spectral efficiency and the energy efficiency can be considered as two opposite performance metrics and it is difficult to improve them at the same time.

In order to find a balance between the spectral efficiency and the energy efficiency, we have proposed a dual-stream phased modulated OFDM transceiver structure in Ref. [7]. Two independent CE-OFDM signals are multiplexed at the transmitter and are joint demodulated at the receiver. The spectral efficiency of this structure is almost same as OFDM. The transceiver structure maintains the PAPR less than 3 dB while does not need complexity signal processing method. Thus, it can be considered as a quasi-constant envelope OFDM system. Since the phase modulation is used, it has been proved that the bit error rate (BER) and the PAPR are related to the modulation index. It should be noted that the spectral efficiency will increase while the energy efficiency will decrease with the increasing of the modulation index. Since a typical satellite communication system contains various types of services, the problem is how to choose the modulation index of each user to get a balance between the spectral efficiency and the energy efficiency and then obtain a maximum system performance.

In this paper, we consider the spectral efficiency and the energy efficiency as two factors and propose a system performance model. The relationship between these two factor and the modulation index are analyzed in this paper. Thus, the relationship between the system performance and the modulation index is established. The proposed modulation index selection strategy is to choose the appropriate modulation index to maximize the system performance. The system performances with different signal-to-noise ratio (SNR) conditions and modulation indices are given through simulation. The optimum modulation indices for different SNR range are obtained according to the simulation results.

The rest of this paper is organized as follows. Section 2 describes the quasi-constant envelope OFDM satellite system model. Section 3 presents the proposed modulation index selection strategy. Section 4 shows the simulation results and analysis. Section 5 concludes the paper.

2 System Model

The system proposed in Ref. [7] is used in this paper, which is shown in Fig. 1. The brief review of the transmission process is given in this section. Capital letters represent the frequency-domain and small letters represent the time-domain. Superscript i is used to distinguish the two streams, where $i = 1, 2$. Assumed that there are N modulated symbols in each stream and they are represented by S_k^i, where $k = 1, 2, \ldots, N$. In order to obtain a discrete-time real-valued version of the OFDM signal, and inverse discrete Fourier transform (IDFT) module is used and the input of the IDFT module should be arranged as a conjugate symmetry structure. Therefore, the input of the IDFT is represented by S_m^i and it can be expressed as

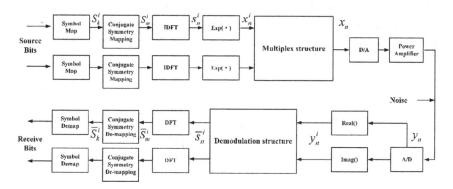

Fig. 1. Quasi-constant envelope OFDM system model

$$S_m^i = [0, S_1^i, S_2^i, \ldots, 0_{1 \times N_{zp}}, 0, S_N^{i*}, \ldots, S_2^{i*}, S_1^{i*}] \quad (1)$$

where $m = 0, 1, \ldots, N_{idft} - 1$, S_k^{i*} is the conjugate form of S_k^i and $0_{1 \times N_{zp}}$ comprises N_{zp} length zero elements. The oversampling factor is defined as $K = N_{idft}/(N_{idft} - N_{zp})$. The output of IDFT can be expressed as

$$s_n^i = \frac{1}{\sqrt{N_{idft}}} \sum_{m=0}^{N_{idft}-1} S_m^i \exp(j2\pi mn/N_{idft}) \quad (2)$$

where $n = 0, 1, \ldots, N_{idft} - 1$.

Due to the feature of OFDM signals [5], the PAPR of s_n^i is high and the value of PAPR is related to the subcarrier numbers. In order to reduce the PAPR, the phase modulator is used to modulate the OFDM signal to the phase of a constant envelope signal. The output of the phase modulator can be expressed as

$$x_n^i = A \exp(j2\pi h s_n^i) \quad (3)$$

where A is the magnitude and $2\pi h$ is the modulation index. The two streams are multiplexed through the multiplex structure and the baseband signal of quasi-constant envelope satellite system can be expressed as

$$x_n = x_n^1 - j \cdot x_n^2 \tag{4}$$

The signal power of x_n can be calculated according to (4) and it can be expressed as

$$|x_n| = 2A^2 + 2A^2 \sin(2\pi h s_n^2 - 2\pi h s_n^1) \tag{5}$$

The signal power is normalized to 1 in this paper. Due to the existence of sine function, the statistical PAPR of the baseband signal is 3 dB. The actual PAPR is influenced by the value of modulation index. Figure 2 presents the power associated to two quasi-constant envelope OFDM signals with $N = 31$, $K = 8$, $A = \sqrt{2}/2$ and $2\pi h = 0.5$, $2\pi h = 0.7$. It can be seen that the PAPR is increasing with the increasing of the modulation index.

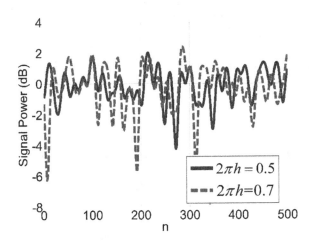

Fig. 2. Signal power of quasi-constant envelope OFDM signal.

Thus, the PAPR can be adjusted by changing the modulation index. Since the nonlinear PA is used, the IBO is needed to avoid the nonlinear signal distortion. Thus, the energy efficiency is related to the PAPR. The detail analysis between energy efficiency and PAPR is given in Sect. 3.

As the aim of this paper is to explain the proposed modulation index selection strategy, the propagation environment is assumed as additive white Gaussian noise (AWGN) channel. The received signal can be expressed as

$$y_n = x_n + w_n \tag{6}$$

where w_n is the complex-valued Gaussian noise with as power spectral density $\Psi_n(f) = N_0$ over the effective bandwidth of x_n [8]. The signal is separated to two streams by taking the imaginary part and the real part, respectively. The two streams at the receiver can be expressed as

$$
\begin{aligned}
y_n^1 &= A[\sin(2\pi h s_n^1) - \cos(2\pi h s_n^2)] + \mathrm{Im}(w_n) \\
y_n^2 &= A[\cos(2\pi h s_n^1) + \sin(2\pi h s_n^2)] + \mathrm{Re}(w_n)
\end{aligned}
\tag{7}
$$

where $\mathrm{Im}(\cdot)$ and $\mathrm{Re}(\cdot)$ represent the real part and the imaginary part of a complex value. The Taylor series expansion [8] is used and y_n^i can be rewritten as

$$
\begin{aligned}
y_n^1 &= A\left[\sum_{l=0}^{\infty} \frac{(-1)^l}{(2l+1)!}(2\pi h s_n^1)^{2l+1} - \sum_{l=0}^{\infty} \frac{(-1)^l}{(2l)!}(2\pi h s_n^2)^{2l}\right] + \mathrm{Im}(w_n) \\
y_n^1 &= A\left[\sum_{l=0}^{\infty} \frac{(-1)^l}{(2l)!}(2\pi h s_n^1)^{2l} + \sum_{l=0}^{\infty} \frac{(-1)^l}{(2l+1)!}(2\pi h s_n^2)^{2l+1}\right] + \mathrm{Re}(w_n)
\end{aligned}
\tag{8}
$$

Then the s_n^i can be obtained by joint demodulation. For example, under the condition of $2\pi h < 0.5$, it has been proved in Ref. [9] that the contribution of the cubic or higher terms of $2\pi h s_n^i$ can be treated as noise. Thus, the $\bar{s}_n^i$ after the demodulation structure can be expressed as

$$
\bar{s}_n^1 = \frac{y_n^1 + A}{2\pi A h}
\tag{9}
$$

$$
\bar{s}_n^2 = \frac{y_n^2 - A}{2\pi A h}
\tag{10}
$$

Therefore, the BER performance is influenced by the modulation index. In this paper, the energy efficiency and the spectral efficiency are considered as two factors which influence the system performance. Since the SNR conditions and the transmission rate demands of users are different, the modulation index should be chosen appropriately to maximize system performance.

3 Modulation Index Selection Strategy

3.1 Energy Efficiency

In order to determine the impact of the PAPR, the solid-state power amplifier (SSSPA) model [10] is assumed in this paper. The SSPA model can be expressed as

$$
x_{out}(t) = \frac{G x_{in}(t)}{[1 + (\frac{|x_{in}(t)|}{A_{sat}})^{2p}]^{\frac{1}{2p}}}
\tag{11}
$$

where $x_{in}(t)$ is the input signal, G is the amplifier gain, A_{sat} is the input saturation level, and p represents the AM/AM sharpness of the saturation region. The IBO is defined as Ref. [11].

$$IBO_{2\pi h} = \frac{A_{sat}^2}{E\{|x_{in}(t)|^2\}} \qquad (12)$$

The PA efficiency is used to quantify the relationship between the energy efficiency and the IBO. The theoretical efficiency of a Class A power amplifier [12] is used and it can be expressed as

$$\eta_{2\pi h} = \frac{1}{2} \frac{1}{IBO_{2\pi h}} \times 100\% \qquad (13)$$

As the IBO is commonly set to be equal to the PAPR, the larger modulation index will further reduce the energy efficiency. The means that more energy is needed to spend for sending the information. The energy spent on each frame is defined as $J_{2\pi h}$, and it can be expressed as

$$J_{2\pi h} = \frac{P_s \cdot T}{\eta_{2\pi h}} \qquad (14)$$

where P_s is the average signal power and T is the frame length.

3.2   Spectral Efficiency

In this paper, the spectral efficiency is defined as the correct bits of each frame divided by the transmission interval and system bandwidth. As the transmission interval is the reciprocal of the subcarrier spacing, the spectral efficiency is equal to the correct bits of each frame divided by the subcarrier numbers. The spectral efficiency can be expressed as

$$\begin{aligned} SE_{2\pi h} &= \frac{N_{bit} \cdot (1 - BER_{2\pi h})}{T \cdot N_{sub} \cdot f_s} \\ &= \frac{N_{bit}(1 - BER_{2\pi h})}{N_{sub}} \end{aligned} \qquad (15)$$

where N_{bit} represents the transmission bits of each frame, $BER_{2\pi h}$ represents the BER value with a modulation index $2\pi h$, N_{sub} is the number of subcarriers which is equal to $(2N + 2)$ and f_s is the subcarrier spacing. Thus, the unit of $SE_{2\pi h}$ is $bit/s/Hz$. Since the BER performance is increasing with the increasing of modulation index, it is apparently that the spectral efficiency is improved at the same time.

3.3   Proposed Modulation Index Selection Strategy

The proposed modulation index selection strategy is expressed as

$$E = \max_{2\pi h} \frac{SE_{2\pi h}}{J_{2\pi h}}$$

$$= \max_{2\pi h} \frac{N_{bit} \cdot (1 - BER_{2\pi h})}{2 \cdot IBO_{2\pi h} \cdot N_{sub} \cdot P_s \cdot T} \times 100\% \tag{16}$$

$$subject\ to\ BER_{2\pi h} \leq BER_{threshold}$$

where E represents the system performance and its unit is $bit/s/Hz/J$. Since it is meaningless to communicate with a high BER, the $BER_{threshold}$ represents the BER restriction condition. The value of $BER_{threshold}$ can be adjusted according to the type of services. The aim is to select the appropriate modulation index which can maximize the E and satisfy the BER restriction condition. In this paper, the relationship between $SE_{2\pi h}$, $\eta_{2\pi h}$ and SNR are obtained through simulation. The performance curves between E and SNR are calculated according to the simulation results.

4   Simulation Results and Analysis

The simulation parameters are shown in Table 1. The amplifier gain G can be considered as a linear power amplification factor. As it does not cause nonlinear distortion, the value of G will not affect the SNR and BER performance. Thus, G is set as 1 in the simulation. 10^7 bits are sent for each E_b/N_0 and the quadrature phase shift keying (QPSK) is the modulation method. As the number of modulation symbols of dual-stream is 2 N, the transmitted bits for each frame is 4 N. The BER threshold is set as 10^{-2} in our simulation. The PAPR performances, the BER performances and the spectral performances are

Table 1. Simulation parameters

Modulation mode	S_k^i	QPSK
Number of symbols per stream	N	31
Number of subcarriers	N_{sub}	64
IDFT points	N_{idft}	1024
Oversampling factor	K	16
Magnitude of signal	A	$\sqrt{2}/2$
Average power of signal	P_s	0 dBw
Frame length	T	64us
Modulation index	$2\pi h$	(0:0.1:1)
Transmission bits for each frame	N_{bit}	124
Amplifier gain	G	1
Input saturation level	A_{sat}	0 dB
AM/AM sharpness	p	3
BER threshold	$BER_{threshold}$	10^{-2}
Target PAPR of OFDM CFS	$PAPR_{Target}$	3 dB
Iteration times	$N_{Iteration}$	16

compared with the OFDM clipping and filtering structure (OFDM CFS) proposed in Ref. [13]. For OFDM CFS, it is assumed that the target PAPR is 3 dB and the iteration times are 16.

The PAPR performances are shown in Fig. 3. It can be seen that the PAPR is getting larger with the increasing of the modulation index. Furthermore, The PAPR of quasi-constant envelope OFDM system is less than the OFDM CFS system. The IBO is set to be equal to the PAPR in the simulation. Therefore, the IBO and the related energy efficiencies of quasi-constant envelope OFDM system under the condition of different modulation index are shown in Table 2. The IBO of OFDM CFS system is 3.5 and the related energy efficiency is 22.33%.

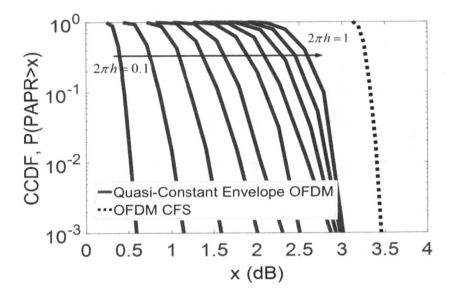

Fig. 3. PAPR performances of quasi-constant envelope OFDM and OFDM CFS system

Table 2. The IBO and the related energy efficiencies

Modulation index	IBO	Energy efficiency
0.1	0.6	43.55%
0.2	1.2	37.93%
0.3	1.6	34.59%
0.4	2.0	31.55%
0.5	2.3	29.44%
0.6	2.5	28.12%
0.7	2.87	25.82%
0.8	2.93	25.74%
0.9	2.98	25.18%
1.0	3	25.06%

The BER performances with different modulation indices are shown in Fig. 4. The influence of nonlinear PA is taken into account. It can be seen that the BER performances are improved with the increasing of the modulation index.

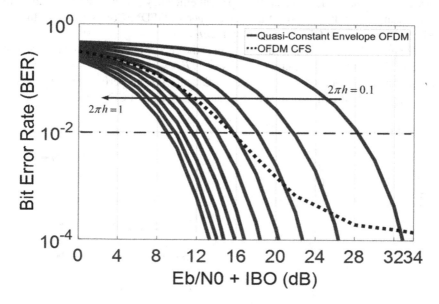

Fig. 4. BER performances of quasi-constant envelope OFDM and OFDM CF system

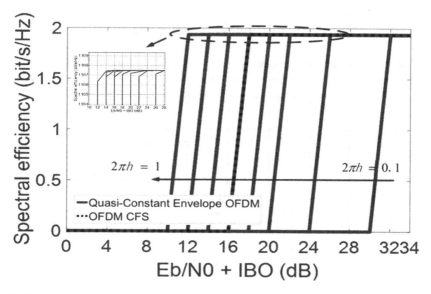

Fig. 5. Spectral performances of quasi-constant envelope OFDM and OFDM CFS system

The spectral efficiencies with different modulation indices can be calculated according to the BER performances. Since the BER restriction condition is determined, if the $BER_{2\pi h} > BER_{threshold}$, the communication is meaning less and the spectral efficiency can be considered as 0. Thus, the spectral efficiency performances are shown in Fig. 5. It can be seen that the spectral performance of OFDM CFS is closed to that of quasi-constant envelope OFDM system with $2\pi h = 0.5$.

The value of E can be calculated according to the energy efficiencies and the spectral efficiency. Figure 6 shows the curves of E versus $E_b/N_0 + IBO$. Since the energy efficiency is taken into account, it can be seen that the system performances of quasi-constant envelope OFDM system is better than that of OFDM CFS system.

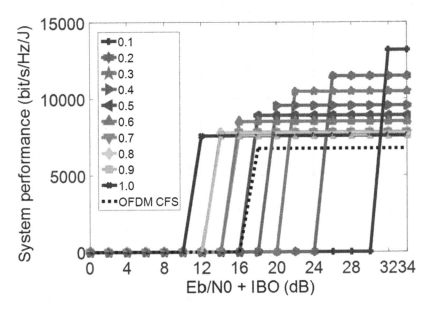

Fig. 6. System performance of quasi-constant envelope OFDM and OFDM CFS system

Therefore, the appropriate modulation indices for different SNR range are shown as Table 3.

Table 3. The selected modulation index

$E_b/N_0 + IBO$	Modulation index
10–12	1
12–14	0.8
14–16	0.6
16–18	0.5
18–20	0.4
20–24	0.3
24–32	0.2
32–34	0.1

It should be noted that the performance enhanced scheme, such as the Low-density Parity-check (LDPC) coding scheme, are not added in our simulation. Thus, the system performance can be further improved and the proposed strategy is suitable for all SNR conditions.

5   Conclusion and Future Work

The paper is focused on the modulation index selection strategy for quasi-constant envelope OFDM satellite system. The system performance model is proposed and the energy efficiency and the spectral efficiency are considered as two factors to evaluate the system performance. The relationship curves between these factors and Eb/N0 under different system parameter assumptions can be obtained by simulation. Since the selected modulation indices for different SNR range are calculated according to the simulation results, in the future, our main focus is on the derivation of the close form solution.

Acknowledgement. This work is supported by the National Natural Science Foundation of China (No. 91438114) and the Science and Technology on Information Transmission and Dissemination in Communication Networks Laboratory (KX152600019/ITU-U15010).

References

1. Murota, K., Hirade, K.: GMSK modulation for digital mobile radio telephony. IEEE Trans. Commun. **29**(7), 1044–1050 (1981)
2. Guo, Y., Feher, K.: A new FQPSK modem/radio architecture for PCS and mobile satellite communications. IEEE J. Sel. Areas Commun. **13**(2), 345–353 (1995)
3. Thompson, S.C., Ahmed, A.U., Proakis, J.G., Zeidler, J.R., Geile, M.J.: Constant Envelope OFDM. IEEE Trans. Commun. **56**(8), 1300–1312 (2008)
4. Weinstein, S.B., Ebert, P.M.: Data transmission by frequency division multiplexing using the discrete Fourier transform. IEEE Trans. Commun. Technol. **19**(5), 628–634 (1971)
5. Imai, H., Ochiai, H.: On the distribution of the peak-to-average power ratio in OFDM signals. IEEE Trans. Commun. **49**(2), 282–289 (2001)
6. Lee, J., Han, S.: An overview of peak-to-average power ratio reduction techniques for multicarrier transmission. IEEE Trans. Wireless Commun. **12**(2), 56–65 (2005)
7. Wang, C., Cui, G.F., Wang, W.D.: Dual-stream transceiver structure with single antenna for phase modulated OFDM. IEEE Commun. Lett. **20**(9), 1756–1759 (2016)
8. Proakis, J.G.: Digital Communications. McGraw Hill, USA (2001)
9. Ahmed, A.U., Zeidler, J.R.: Novel low complexity receivers for constant envelope OFDM. IEEE Trans. Signal Process. **63**(17), 4572–4582 (2015)
10. Thompson, S.C.: Constant Envelope OFDM Phase Modulation. University of California, San Diego (2005)
11. Ochiai, H.: Performance analysis of peak power and band limited OFDM system with linear scaling. IEEE Trans. Wireless Commun. **2**(5), 1055–1065 (2003)
12. Ochiai, H.: Power efficiency comparison of OFDM and single-carrier signals. In: Vehicular Technology Conference, Vancouver, pp. 899–903 (2002)
13. Armstrong, J.: Peak-to-average power reduction for OFDM by clipping and frequency domain filtering. Electron. Lett. **38**(5), 246–247 (2002)

Wide-Angle Scanning Phased Array Antenna for the Satellite Communication Application

Chunmei Liu and Shaoqiu Xiao[(⊠)]

Institute of Applied Physics,
University of Electronic Science and Technology of China, Chengdu, China
xiaoshaoqiu@uestc.edu.cn

Abstract. A novel microstrip magnetic dipole antenna with wide beam-width is proposed in this paper. It is working as a cavity backed slot antenna. The half power beam-width (HPBW) is 150° in the E-plane, and 85° in the H-plane. Its low profile of 0.03 λ_0 (λ_0 is wavelength of operating center frequency) and compact size of 0.31 λ_0 * 0.32 λ_0 make it suitable for the space communication systems. Due to the good performance of the antenna element, two linear phased arrays have been constructed, E-plane and H-plane phased arrays. The main lobe can scan from −76° to 76° for the E-plane phased array, and from −65° to 65° for the H-plane phased array with about a 3 dB gain fluctuation. Moreover, a planar phased array has been constructed. The scanning range is from −55° to 55° in all azimuth planes.

Keywords: Microstrip magnetic dipole antenna · Wide-angle scanning phased array · Low profile · Low gain fluctuation

1 Introduction

Space information network (SIN) is a system, carried by satellite, stratospheric airship, and unmanned aerial vehicle (UAV) to receive and transmit the space information, then process, which is the core of modern wireless communication systems. Due to the fast development of wireless communication, more wireless devices will connect to the wireless communication network, which will cause information congestion. Moreover, the available frequency bands are limited. The antenna, as a front-end equipment of wireless communication system, determines the system performance. Therefore, the wide bandwidth and wide radiation coverage of antenna are required to improve the efficiency of frequency and space spectrum. Many methods have been proposed to improve the property of antennas. In many cases, only the wide coverage is urgently demanded. For the specific situation of SIN, low profile and small bulk should be considered first. Moreover, due to the telecommunication, high-power is urgently demanded. The planar microstrip array antenna is a good candidate for space communication system due to the low profile and low cost. However, the radiation coverage of array antenna is limited by the half power beam-width (HPBW) of antenna element [1–4].

Many techniques, to enlarge the HPBW of the microstrip antennas, have been developed in recent years. Such as pattern-reconfigurable antennas, microstrip

© Springer Nature Singapore Pte Ltd. 2017
Q. Yu (Ed.): SINC 2016, CCIS 688, pp. 162–171, 2017.
DOI: 10.1007/978-981-10-4403-8_15

magnetic antennas [5–12]. However, these methods have a disadvantages of complicated design or large size in one dimension, which are not suitable for the two dimensions scanning phased arrays applied for the SIN systems.

In this paper, a novel microstrip magnetic dipole antenna is proposed. It has an omnidirectional radiation pattern in the E-plane and wider beam-width in the H-plane on the upper half space. Two linear phased arrays along the E-plane and H-plane have been arranged, respectively. The main lobe can scan from −76° to 76° for E-plane phased array, and −64° to 64° for H-plane phased array with low fluctuation. A planar phased array has also been constructed, and the simulated result shows that the main lobe can scan from −55° to 55° in all azimuth planes with a low gain fluctuation.

2   Design of Wide Beam-Width Antenna Element

2.1   Structure of the Proposed Antenna

The structure of the proposed antenna is shown in Fig. 1. The size is 0.32 λ_0 * 0.31 λ_0 * 0.03 λ_0, which is appropriate for applying to two dimensions wide-angle scanning phased array application. The proposed antenna is built on a F4BM substrate with thickness of h = 1.5 mm, the relative permittivity of substrate ϵ_r = 2.2. Due to limitation of the process of PCB, the metal walls were implemented by via holes. The parameters of the proposed microstrip magnetic antenna is shown at Table 1.

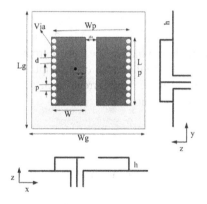

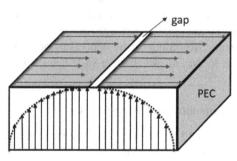

Fig. 1. Structure of the proposed antenna element.

Fig. 2. Photographs of proposed antenna element. (a) top view. (b) back view.

This antenna can be analyzed by the cavity model, similar to the microstrip patch antenna. The field distribution within the cavity can be obtained by using the vector potential approach. Then the dominant mode with the lowest resonance can be determined. Due to the homogeneous wave equation and boundary condition, the electric field component can be obtained from (1), (2), (3) and (4) [4].

Fig. 3. Photographs of proposed antenna element. (a) top view of the antenna. (b) bottom view of the antenna.

Table 1. Parameters of proposed antenna

Symbol (mm)	Value
ε_r	2.2
Lg	60
Wg	60
Lp	17
Wp	7.8
h	1.5
dis	1.2
d	0.6
p	0.9

$$\nabla^2 A_z + k^2 A_z = 0 \tag{1}$$

$$E_z = -j\frac{1}{\omega\mu\varepsilon}\left(\frac{\partial^2}{\partial x^2} + k^2\right)A_z \tag{2}$$

$$E_z(x = 0, x = W_p) = 0 \tag{3}$$

$$E_z = E_0 \sin\left(\frac{\pi}{W_p}x\right) \tag{4}$$

The current distribution of the patch can be obtained, as shown in Fig. 2, There must be a displacement current at the slot according to the continuity of electric current. This displacement current is the main radiation source due to the radiation principle of antennas. According to the equivalent principle, this displacement current equal to a magnetic current, which have wide beam-width in the E-plane.

2.2   Simulated and Measure Results

HFSS is used to simulate this antenna. It has been fabricated and measured. The photographs are shown in Fig. 3.

The reflection coefficient is tested by Vector Network Analyzer E8361A. The simulated and measured reflection coefficients are shown in Fig. 4. This figure shows the measured result is consistent with the simulated one. The proposed antenna works at the band from 5.6 GHz to 5.9 GHz. The radiation patterns have been measured at an anechoic chamber, shown in Fig. 5. The HPBW is about 150° in the E-plane, and 85° in the H-plane.

The comparison between the proposed antenna and conventional half-wavelength patch antenna has been finished, shown in Fig. 6. Obviously, the HPBW of the proposed antenna is wider than the conventional half-wavelength patch antenna.

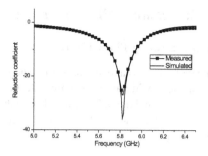

Fig. 4. Reflection coefficient of single antenna

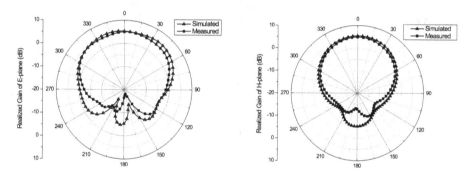

Fig. 5. Radiation of parallel magnetic dipole with a finite ground. (a) E-plane. (b) H-plane.

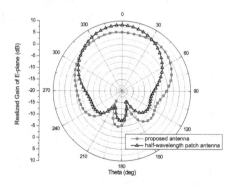

Fig. 6. Radiation of propose magnetic dipole and half-wavelength patch antenna with a finite ground.

3   Linear Wide-Angle Scanning Phased Arrays

3.1   E-Plane Wide-Angle Scanning Phased Array

Due to the good performance in the E-plane, the proposed antenna element is applied to a E-plane phased array to achieve wide-angle scanning. A 9-elements E-plane phased array is constructed and simulated by HFSS. It has also been fabricated. The parameters of the phased array are shown at Table 2. The distance between elements has been optimized, which is about $0.4\lambda_0$.

Table 2. Parameters of phased array antenna

Symbol (mm)	E-plane Phased array	H-plane phased array	Plane phased array
ε_r	2.2	2.2	2.2
Lg	80	250	400
Wg	250	60	400
Lp	16	16	16
Wp	7.7	7.85	7.8
h	1.5	1.5	1.5
dis	1.2	1.2	1.2
d	0.6	0.6	0.6
p	0.9	0.9	0.9
disx	0	19	19
disy	21	0	21

The photograph is shown in Fig. 7. The reflection coefficients have been tested by Vector Network Analyzer E8361A, shown in Fig. 7. It can be seen that the reflection coefficients of the center elements are different with the edge elements due to the edge effect and the coupling effect. As the result shows, measured working band of central elements are bigger than the isolated one, from 5.6 GHz to 6.15 GHz.

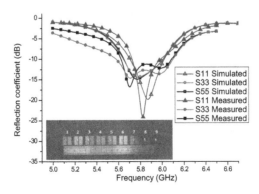

Fig. 7.   Reflection coefficient of E-plane phased array

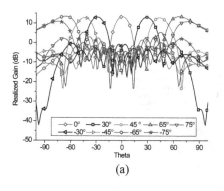
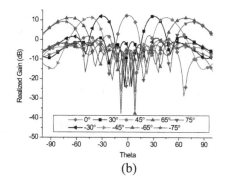

(a)   (b)

Fig. 8. Scanning results in the E-plane (yoz) at 5.8 GHz, (a) simulated result, (b) measured result.

In this case, there is only phase controlling. From the simulated result, it can be seen that the main lobe can scan from $-76°$ to $76°$ in E-plane with a 3 dB gain fluctuation, which is shown in Fig. 8(a). Corresponding, the measured result shows a good agreement with simulation, shown in Fig. 8(b). The information of other angles is shown in Table 3. From the results, there is no blind point and granting lobe at the whole upper half space due to the small distance between antenna elements. Although the distance between elements is small, no catastrophic distortion emerges at array gain because the mutual coupling has been reduced by the via holes (Table 4).

Table 3. Radiation pattern information of scanning performance at 5.8GHZ in the E-Plane

θ_0	Realized (dB)		3 dB beam coverage	SLL (dB)
	Simulated	Measured		
0°	13.4	12.2	–8°–8°	–13.2
10°	13.3	12.4	2°–18°	–13.1
20°	13.4	12.2	13°–27°	–13.3
30°	13.0	11.6	22°–39°	–13.5
40°	12.7	11.2	30°–50°	–13.7
50°	12.8	10.5	40°–60°	–14.8
60°	13.3	10.1	52°–75°	–13.6
70°	12.8	9.8	58°–88°	–7.4
80°	9.5	6.1	68°–95°	–5

Table 4. Radiation pattern information of scanning performance at 5.8GHZ in the H-plane

θ_0	Realiazed gain (dB)		3 dB beam coverage	SLL (dB)
	Simulated	Measured		
0°	11.2	9.4	–8°–8°	–12.4
10°	11.4	9.7	2°–18°	–12.9
20°	11.7	9.9	13°–27°	–14.3
30°	12.4	10.8	19°–41°	–9.9
40°	12.5	11.3	32°–50°	–13.1
50°	11.5	11.5	42°–53°	–10.2
60°	9.3	9.7	51°–74°	–6.9
65°	6.8	6.1	56°–80°	–3.8
70°	2.8	3.4	63°–85°	–1

3.2 H-Plane Wide-Angle Scanning Phased Array

Due to the compact size and relatively wider HPBW in the H-plane, this antenna can also be applied to a H-plane scanning phased array without grating lobe. A 9-elements H-plane phased array is constructed, as shown in Fig. 9. For the H-plane phased array, the coupling is stronger compared with E-plane due the opened sides. Fortunately, the

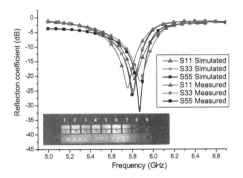

Fig. 9. Reflection coefficient of H-plane phased array

active radiation pattern can be controlled by adjusting the coupling based on the antenna aperture synthesis [4]. In this situation, the small size of antenna element makes it possible to control the coupling easily. The distance between elements along the H-plane has been optimized, which is 0.37 λ. A quasi-sector radiation pattern in the H-plane is obtained. Meanwhile, the scanning blind point can be eliminated due to the tight arrangement.

Some parameters have been adjusted to make array work at the same band with the E-plane array, as shown in Table 2. The simulation and fabrication have been done. It works at the band from 5.6 GHz to 5.9 GHz. Measured result is similar to the simulation. It is shown in Fig. 9.

The scanning has been finished by changing the phased progression. The scanning range is smaller than the E-plane due to the narrower beam-width of element in the H-plane. The main lobe can scan from $-64°$ to $64°$ with a 3 dB gain fluctuation. The measured result shows a good agreement with the simulated result. It is shown in Fig. 10.

4 2D Phased Array

4.1 Structure of the 2D Phased Array

This magnetic dipole antenna has compact size which is very suitable for a two dimensions wide-angle scanning phased array. Due to the good performance of wide-angle scanning phased arrays at two planes, we have arranged a two dimensions phased array. The structure of the phased array is shown in Fig. 11, and the parameters is shown at Table 2.

4.2 Simulated Result

From the simulated results, this phased array behaves a good performance in all azimuth planes. The main lobe can scan from $-55°$ to $55°$ in H-plane, $-55°$ to $55°$ in E-plane, $-60°$ to $60°$ in D-plane, with 3 dB gain fluctuation. The cross polarization has

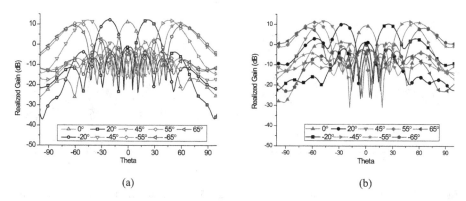

(a)   (b)

Fig. 10. Scanning results in the H-plane (yoz) at 5.8 GHz (a) simulated result, (b) measured result

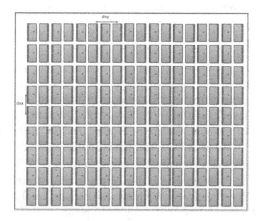

Fig. 11. Structure of plane phased array

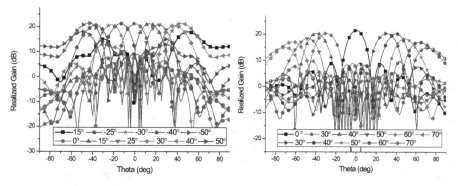

Fig. 12. Scanning results in the H-plane (yoz) at 5.8 GHz

Fig. 13. Scanning results in the E-plane (xoz) at 5.8 GHz

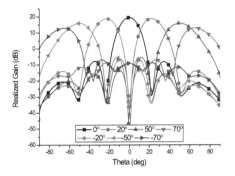

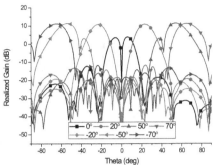

Fig. 14. Scanning results in the D-plane at 5.8 GHz

Fig. 15. Cross polarization of the D-plane scanning at 5.8 GHz

been shown in Fig. 15. When the main lobe scan to large angles, the gain decreases fast because the large cross polarization (Figs. 12, 13 and 14).

5 Conclusion

In this paper, a novel microstrip magnetic antenna is proposed. It has a low profile and compact size, which is very suitable for SIN systems. The antenna element is resonant at about 5.8 GHz. The HPBW is 150° in the E-plane, and 85° in the H-plane. Based on the good performance of the antenna element, it is applied to the E- and H-plane phased arrays. The results show that the main lobe can scan from $-76°$ to $76°$ in the E-plane phased array and from $-64°$ to $64°$ in the H-plane phased array with a low gain fluctuation about 3 dB, respectively. We have also construct a 9 * 9 elements phased array. Considered the cost of phased array, this plane phased array has not been fabricated. From the simulated result, it can been seen that the main lobe can scan from $-55°$ to $55°$ in all azimuth planes.

References

1. Mailloux, R.J.: Phased Array Antenna Handbook, 2nd edn. Artech House, Norwood (2005)
2. Garg, R., Bhartia, P., Bahl, I., Ittipibooh, A.: Microstrip Antenna Design Handbook. Artech House, Norwood (2001)
3. Schmidt, F.W.: Low-cost microstrip phased array antenna for use in mobile satellite telephone communication service. In: Proceedings of the IEEE Antennas Propagation Society International Symposium, vol. 25, pp. 1152–1155 (1987)
4. Harrington, R.F.: Time-Harmonic Electromagnetic Field
5. Bai, Y.Y., Xiao, S.Q., Tang, M.C., Ding, Z.F., Wang, B.-Z.: Wideangle scanning phased array with pattern reconfigurable elements. IEEE Trans. Antennas Propag. **59**(11), 4071–4076 (2011)

6. Bai, Y.Y., Xiao, S.Q., Wang, B.-Z., Ding, Z.F.: Applying weighted thinned linear array and pattern reconfigurable element to extend pattern scanning range of millimeter wave microstrip phased array. J. Infrared Millim. Terahertz Waves **31**(1), 1–6 (2010)
7. Ding, X., Wang, B.-Z., He, G.-Q.: Research on a millimeter-wave phased array with wide-angle scanning performance. IEEE Trans. Antennas Propag. **61**(10), 5319–5324 (2013)
8. Xiao, S., Zheng, C., Li, M., Xiong, J.: Varactor-loaded pattern reconfigurable array for wide-angle scanning with low gain fluctuation. IEEE Trans. Antennas Propag. **63**(5), 2364–2369 (2015)
9. Wang, R., Wang, B.Z., Ding, X., Yang, X.S.: Planar phased array with wide-angle scanning performance based on image theory. IEEE Trans. Antennas Propag. **63**(9), 3908–3917 (2015)
10. Li, M., Xiao, S., Bai, Y.-Y., Wang, B.-Z.: An ultrathin and broadband radar absorber using resistive FSS. IEEE Antenna Wireless Propag. Lett. **11**, 748–751 (2012)
11. Yan, L., Hong, W., Hua, G., Chen, J., Wu, K., Cui, T.J.: Simulation and experiment on SIW slot array antennas. IEEE Microw. Wireless Compon. Lett. **14**(9), 446–448 (2004)
12. Wen, Y.-Q., Wang, B.-Z., Ding, X.: A wide-angle scanning and low sidelobe level microstrip phased array based on genetic algorithm optimization. IEEE Trans. Antennas Propag. **64**(2), 805–810 (2016)

A Movable Spot-Beam Scheduling Optimization Algorithm for Satellite Communication System with Resistance to Rain Attenuation

Houtian Wang[1(✉)], Xingpei Lu[2], Dong Chen[1], Yufei Shen[1], Ying Tao[1], Zihe Gao[1], and Guoli Wen[2]

[1] China Academy of Space Technology (CAST), Beijing, China
1357953688@qq.com
[2] Beijing University of Posts and Telecommunications, Beijing, China

Abstract. The movable spot-beam antenna of satellite can adjust its pointing flexibly, so it can meet the on-demand coverage need for the ground area. On the other hand, the Ka band is more and more utilized in the satellite communication system with the increasing demand of multi-media data business from users. However, the Ka band is susceptible to rain attenuation. In this paper, a kind of movable spot-beam schedule algorithm for resisting rain attenuation in satellite networks (MSBSA-RRA) is proposed for the adverse effects of rain attenuation on the communication performance of Ka-band. MSBSA-RRA establishes the optimal model of spot-beam antenna scheduling by using the optimization theory. Two kinds of users are set up in this paper, which are special users and ordinary users. The priority for special users is higher and its service quality needs to be guaranteed underlying. The attenuation margin factor is introduced in the optimization model, so the spot-beam antenna can give priority to special users under rainfall conditions. The optimization model is solved by classifying the spot-beam scheduling problem as the maximum weight perfect matching problem of bipartite graph. The performance of MSBSA-RRA is simulated and analyzed in this paper. Simulation results show that MSBSA-RRA can effectively reduce the drop call rate in the rainfall condition. Meanwhile, the influence of the attenuation margin factor on the performance of special users and ordinary users is also analyzed in this paper. It can be seen from the simulation results that the value of attenuation has a very important influence on the performance of MSBSA-RRA.

Keywords: Satellite communication system · Movable spot-beam scheduling optimization · Rain attenuation · The maximum weight perfect matching problem of bipartite graph

1 Introduction

The demand of communication satellite payload is growing. The movable spot-beam antenna of satellite can adjust its pointing flexibly, so it can meet the on-demand coverage need for the ground area. Moreover, the movable spot-beam antenna has the advantages of simple structure and high precision. So it is more and more applied to all

© Springer Nature Singapore Pte Ltd. 2017
Q. Yu (Ed.): SINC 2016, CCIS 688, pp. 172–181, 2017.
DOI: 10.1007/978-981-10-4403-8_16

kinds of spacecraft. When the satellite antenna needs to cover multiple targets in a specific area, antenna pointing strategy needs to be developed in order to optimize the resource utilization of movable spot-beam antenna. That is, optimization algorithm should be developed to determine the location the antenna points to.

In recent years, domestic and foreign scholars have carried out some research on the pointing problem of the movable spot-beam antenna [1–4]. A low complexity heuristic based on graph theory is proposed in [5] for the Multiple-Input Multiple-Output (MIMO) system in order to find a scheduling plan that well balances the offered rates to the served users. In [6], the development of an antenna pointing control system for the operation of satellites orbiting in Low-Earth Orbit is presented. A high-fidelity orbital dynamics model is developed for the prediction of the satellite's position and velocity. In [7], the arithmetic of satellite phased array scanning was deduced and a simplified model for on board phased array scanning was presented. The arithmetic in this paper gave out the formula of computing the planar angle of satellite phased array scanning based on the satellite orbit and attitude, and during the flight the phased array kept pointing to the targeted point. Based on the study of satellite spot beam antenna, an antenna power radiation model is introduced into calculation model of satellite pointing to complete the gain calculation of a pointing target in [8]. The optimization principle is used to analyze the optimization model of the multi-target pointing gain. In [9], a modified algorithm of spot-beam pointing based on iterative search is proposed. Simulation indicates that the proposed algorithm could effectively reduce the moving times of spot-beam antenna while satisfying the requirements of communication. But the existing movable spot-beam scheduling algorithms do not consider the actual channel conditions. When the channel condition is deteriorating, the communication quality of the user can not be guaranteed.

On the other hand, with the development of satellite communication and increasing demand of multi-media data business from users, the Ka band is more and more utilized. At present, O3b company plans to deploy 16 Middle Earth Orbit satellites and every satellite contains 12 movable spot-beam antenna. In order to achieve user's broadband access, Ka band is utilized [10, 11]. Therefore, the movable spot-beam that makes use of Ka band can provide high speed, large bandwidth services as well as on-demand coverage for users. However, the link performance of Ka band satellite communication system is affected by rain attenuation, which leads to the deterioration of the channel conditions.

In this paper, a kind of movable spot-beam schedule algorithm for resisting rain attenuation in satellite networks (MSBSA-RRA) is proposed. Special users and ordinary users are considered, and the priority of special users is higher. In MSBSA-RRA, the optimal model of spot-beam antenna scheduling is established. The attenuation margin factor is introduced for special users. The purpose is to maximize the number of users that the movable spot-beams can cover under the premise of providing quality of service (QoS) for special users underlying. The optimization model is solved by classifying the spot-beam scheduling problem as the maximum weight perfect matching problem of bipartite graph. This paper is structured as follows: the first part is the introduction. In the second part, the scheduling optimization model of MSBSA-RRA is constructed and the execution steps of MSBSA-RRA are given in the third part. The complexity of the algorithm is also analyzed in this part. In the fourth part, the effectiveness of MSBSA-RRA is analyzed by means of simulation. The last part gives the conclusion.

2 The System Model

In this paper, a mobile satellite communication system contains S satellites is considered. Every satellite is equipped with B movable spot-beam antennas and the number of user terminals is U. Users are divided into two types: ordinary users and special users. It is assumed that the number of users that can be covered is Nc after the completion of the movable spot-beam scheduling process. The optimization model established in this paper is as follows:

$$obj \; max(Nc)$$

$$s.t. \begin{cases} \alpha \geq \text{Elevation}_{th} \\ \beta \leq \text{Swing}_{th} \\ \theta \leq \text{Hpbw}_{th} \\ (E_b/N_0)_{special} \geq D_t + F_m \\ (E_b/N_0)_{usual} \geq D_t \end{cases} \qquad (1)$$

And $Nc = \sum\limits_{i=1}^{S} \sum\limits_{j=1}^{B} bc_{ij}$.

bc_{ij} stands for the number of users terminals covered by the jth ($1 \leq j \leq b$) movable spot-beam of the ith ($1 \leq i \leq s$) satellite. α stands for the angle between the horizon and the line connecting the satellite and the user terminal. Elevation_{th} stands for the minimum communication elevation of user terminal. The conditions for the establishment of communication between the satellite and the user terminal are $\alpha \geq \text{Elevation}_{th}$. β stands for the angle between the user-satellite connection and the satellite-gravity connection. Swing_{th} stands for the maximum swing range of a satellite's movable spot-beam antenna. θ stands for the angle the user-satellite connection deviate from the center point of the antenna that covers the user. Hpbw_{th} stands for the half power angle of the movable spot beam antenna. If $\beta \leq \text{Swing}_{th}$ and $\theta \leq \text{Hpbw}_{th}$, then the user can be served by the movable spot beam antenna. $(E_b/N_0)_{special}$ stands for the signal to noise ratio of the downlink obtained by the link budget for special users. D_t stands for the demodulation threshold. F_m stands for the attenuation margin factor reserved for special users. If $(E_b/N_0)_{special} \geq D_t + F_m$, then special users and the satellite can set up communication successfully. $(E_b/N_0)_{usual}$ stands for the signal to noise ratio of the downlink obtained by the link budget for ordinary users. If $(E_b/N_0)_{usual} \geq D_t$, then ordinary users and the satellite can set up communication successfully. The calculation methods of every parameter in (1) are as follows:

Suppose that $\mathbf{O}$ is the center of the earth, $\mathbf{S_1}$ is the satellite, $\mathbf{A_1}$ is the user location, $\mathbf{B_1}$ is the point center of a movable spot beam antenna covering a certain area, and the area contains $\mathbf{A_1}$, $\mathbf{C_1}$ is the point center when the swing angle of the movable spot beam antenna takes the maximum. Figure 1 shows the cases described above.

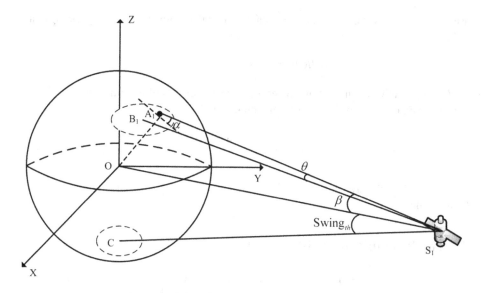

Fig. 1. The pointing schematic of movable spot beam antenna

According to cosine theorem,

$$\alpha = asin(\frac{\overrightarrow{OA_1} * \overrightarrow{A_1S_1}}{\left|\overrightarrow{OA_1}\right| * \left|\overrightarrow{A_1S_1}\right|}) \tag{2}$$

$$\beta = acos(\frac{\overrightarrow{S_1A_1} * \overrightarrow{S_1O}}{\left|\overrightarrow{S_1A_1}\right| * \left|\overrightarrow{S_1O}\right|}) \tag{3}$$

$$\theta = acos(\frac{\overrightarrow{S_1A_1} * \overrightarrow{S_1B_1}}{\left|\overrightarrow{S_1A_1}\right| * \left|\overrightarrow{S_1B_1}\right|}) \tag{4}$$

E_b/N_0 stands for the signal to noise ratio for the user. Link budget is calculated as follows:

$$E_b/N_0 = [C/N]_d - 10\lg R_b + 10\lg(B) \tag{5}$$

$[C/N]_d$ stands for the carrier to noise ratio of the user downlink. R_b stands for the information transmission rate of the user downlink. B stands for the receiver bandwidth. The calculation method of $[C/N]_d$ is shown as follows:

$$[C/N]_d = [EIRP]_s - L_d - \Delta L_d + [G/T]_e - 10\lg(kB) \tag{6}$$

$[EIRP]_s$ stands for equivalent isotropic radiated power, L_d stands for the path loss, ΔL_d stands for the downlink additional loss, $[G/T]_e$ stands for the quality factor of the user terminal, k is the boltzmann constant. The calculation formula of $[EIRP]_s$ is as follows:

$$[EIRP]_s = P_s - L_t + G_t \qquad (7)$$

P_s is the satellite transmitting power, L_t is the feeder loss of the transmitter, G_t is the transmit gain of the movable spot beam antenna and

$$G_t = G_0 * P(\theta) = G_0 * \frac{4J_1^2(\pi\theta D/\lambda)}{(\pi\theta D/\lambda)^2} \qquad (8)$$

Where G_0 is the transmit gain of the antenna's center point. $J_1(x)$ is the first step Bessel function. D is the antenna aperture.

3   The Principle of MSBSA-RRA

In order to solve the optimization model shown in (1), three sets are set up in this paper: Satellite set $\mathbf{S}_{set\neg}$, movable spot beam set $\mathbf{B}_{set}$ and user set $\mathbf{U}_{set}$ (Fig. 2).

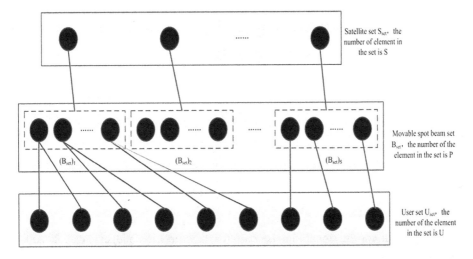

Fig. 2.   Satellite set $S_{set\neg}$, movable spot beam set B_{set} and user set U_{set}

As mentioned in the second part of this paper, there are **B** movable spot beam in each satellite. So, $\mathbf{B}_{set}$ can be divided into **S** sub sets $(\mathbf{B}_{set})_1, (\mathbf{B}_{set})_2, \ldots, (\mathbf{B}_{set})_S$. Multiple users can be covered by a certain movable spot beam and a user can only use one movable spot beam for communication. So solving the optimization model established in this paper is equivalent to get the best matching between the elements in $\mathbf{B}_{set}$ and $\mathbf{U}_{set}$. At present, a lot of combination optimization problems in the field of information science can be transformed into the maximum weight perfect matching problem of bipartite graph [12]. In this paper, the optimization model is solved by the method of the maximum weight perfect matching problem of bipartite graph. The solving process of MSBSA-RRA is as follows:

(1) Polling each satellite node in $\mathbf{S_{set}}$ and get the set $(\mathbf{V_i})_{set}$. Every element in $(\mathbf{V_i})_{set}$ satisfies the condition $\alpha \geq Elevation_{th}$ for the ith satellite. And $\left(V_i\right)_{set} \subseteq U_{set}$.

(2) For the ith satellite in $\mathbf{S_{set}}$, polling each movable spot beam in this satellite. For the jth movable spot beam in the ith satellite, crawling each element in $(\mathbf{V_i})_{set}$. Under the condition that the kth user is taken as the center, MSBSA-RRA calculates the number of users $Cover_{ijk}$ which are in the same beam with user k. The criterion for judging whether the users are covered by the same movable spot beam is as follows:
For the special users, the condition $(E_b/N_0)_{special} \geq D_t + F_m$, $\beta \leq Swing_{th}$ and $\theta \leq Hpbw_{th}$ should be satisfied;
For the ordinary users, the condition $(E_b/N_0)_{usual} \geq D_t$, $\beta \leq Swing_{th}$ and $\theta \leq Hpbw_{th}$ should be satisfied;

(3) Calculate $\max\limits_{k\in(V_i)_{set}} Cover_{ijk}$, which means the jth movable spot beam in the ith satellite points to the latitude and longitude of user k. Then get the set $(\mathbf{C_{ijk}})_{set}$. The elements in $(\mathbf{C_{ijk}})_{set}$ are covered by the same beam with user k.

(4) $(V_i)_{set} = (V_i)_{set} - (C_{ijk})_{set}$. For the (j + 1)th movable spot beam in satellite i, step (3) is repeated until the elements in $\mathbf{B_{set}}$ are crawled or $(V_i)_{set} = \Phi$. Suppose $(SU_i)_{set}$ is the set of users that served by the ith satellite.

(5) For the (j + 1)th satellite, if $(V_{j+1})_{set} \neq \Phi$, then $(V_{j+1})_{set} = (V_i)_{set} - (SU_i)_{set}$. Step (2) is repeated until every user or every satellite is crawled.

The time complexity that calculating the optimal pointing of a certain movable spot beam is $O(|(V_i)_{set}|^2)$. $|(V_i)_{set}|$ stands for the number of elements in set $(\mathbf{V_i})_{set}$. For a satellite communication system with $\mathbf{P}$ movable spot beams, The time complexity of MSBSA-RRA is $O(P \times |(V_i)_{set}|^2)$. So, the algorithm can be solved in polynomial time (Fig. 3).

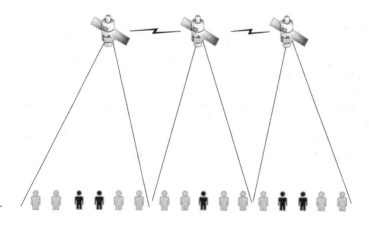

Fig. 3. Simulation scenario

4 Simulation Analyses

4.1 Simulation Scenario

This paper refers to the structure of O3B satellite system so as to provide continuous coverage for global hot spots [13]. Simulation parameters are shown in Table 1.

Table 1. Simulation parameters

Frequency	20 GHz
Antenna aperture	0.5 m
Falf-power angle of the beam	2.5°
Bandwidth	500 MHz
Data rate	36 Mbps
Output power	17.5 dBW
Feeder link loss	4.25 dB
Downlink demodulation threshold	5 dB

Special users and ordinary users are contained in the simulation scenario and these users are randomly distributed in the global hot spots. By setting the rain attenuation value, the performance of MSBSA-RRA is analysis.

4.2 Simulation Analysis

The performance of MSBSA-RRA and the algorithm proposed in [14] is compared in this subsection. In [14], a fast algorithm for optimal satellite spot beam pointing (FA-OSSBP) is proposed. The aim is to maximize the number of users that can be covered as well as improve the implementation efficiency. However, FA-OSSBP does not consider the actual satellite channel condition when the movable spot beam is performed.

This paper draws lessons from realization ideas of rain attenuation model in [15, 16]. The performance of drop call rate is compared when the rain attenuation is 9 dB. The result is shown in Fig. 4.

As is shown in Fig. 4, T represents the run cycle of the medium orbit satellite. In this paper, the performance of the two algorithms in T is analyzed. Under the condition of rain attenuation, the call drop rate is lower when MSBSA-RRA is executed. The reason is that the attenuation margin factor is introduced in MSBSA-RRA, so the movable spot beam can be moved to the location of special users as far as possible under the condition of rainfall. The quality of the received signals of special users can be enhanced and the QoS of special users is better.

Figure 5 shows that when the rain attenuation is 6 dB and 9 dB respectively, the call drop rate of special users with the attenuation margin factor varying from 1 dB to 10 dB.

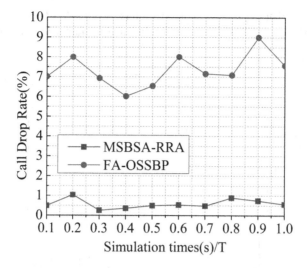

Fig. 4. The performance comparison between MSBSA-RRA and FA-OSSBP

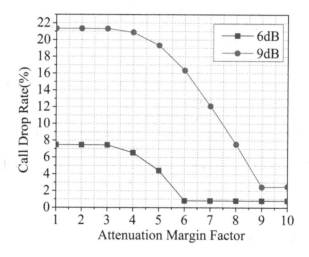

Fig. 5. The call drop rate of special users with the attenuation margin factor varying from 1 dB to 10 dB

As is shown in Fig. 5, when the attenuation margin factor is larger than 3 dB, the call drop rate decreased with the increase of the attenuation margin factor. As the value of the attenuation margin factor is larger than the rain attenuation, the call drop rate tends to be stable. The call drop rate of special users is convergent to 0.7% and 2.2% when the rainfall is 6 dB and 9 dB respectively. When the attenuation margin factor is larger, the beam center tends to point to the location of special users. Therefore, it can provide a greater gain for special users.

Figure 6 shows that when the rain attenuation is 6 dB and 9 dB respectively, the call drop rate of ordinary users with the attenuation margin factor varying from 1 dB to 10 dB.

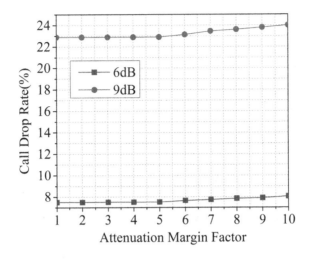

Fig. 6. The call drop rate of ordinary users with the attenuation margin factor varying from 1 dB to 10 dB

As is shown in Fig. 6, with the increase of the attenuation margin factor reserved for special users, the call drop rate of ordinary users presents a rising trend. The reason is that the introduction of the attenuation margin factor makes the beam center tends to point to the location of special users. So it will sacrifice the QoS of some ordinary users. We can conclude from Figs. 5 and 6 that the attenuation margin factor for special users is larger, the ability of special users to resist rain attenuation is stronger. However, the QoS of ordinary users will be sacrificed. The value of the attenuation factor needs to be considered further.

5   Conclusion and Future Work

In this paper, the attenuation margin factor is introduced for special users when special users and ordinary users exist at the same time. The aim is to provide quality of service to special users firstly under the condition of rainfall. On this basis, the movable spot beam scheduling optimization model is constructed. The method of the maximum weight perfect matching problem of bipartite graph is utilized to solve this optimization model. Finally, MSBSA-RRA is compared with FA-OSSBP. Simulation results show that call drop rate of special users is lower when MSBSA-RRA is running. Moreover, the call drop rate of special users and ordinary users with the attenuation margin factor varying is analyzed in this paper. The greater the attenuation margin factor is reserved for special users, the stronger the ability of special users to resist the rain attenuation. But the QoS of some ordinary users will be sacrificed. So the value of the attenuation margin factor is important for MSBSA-RRA and need to be considered further. The proposed algorithm and conclusions in this paper can effectively guide the practical engineering. It has the engineering directive significance for the satellite antenna pointing and covering problem.

Acknowledgments. This work is supported by the National Natural Science Foundation of China (No. 61271281, No. 91438205) and Equipment Development Foundation of China (No. 9140A21010115HT05003).

References

1. Thornton, J.: A low sidelobe asymmetric beam antenna for high altitude platform communications. IEEE Microwave Wirel. Compon. Lett. **14**, 59–61 (2004)
2. Klein, C.: Design of shaped-beam antennas through minimax gain optimization. IEEE Trans. Antennas Propag. **32**, 963–968 (1984)
3. Chang, D.C., Hu, C.N., Hung, C.I., Ho, K.T.: Pattern synthesis of the offset reflector antenna system with less complicated phased array feed. IEEE Trans. Antennas Propag. **42**, 240–245 (1994)
4. Isernia, T., Bucci, O.M., Fiorentino, N.: Shaped beam antenna synthesis problems: feasibility criteria and new strategies. J. Electromagn. Waves Appl. **12**, 103–138 (1998)
5. Boussemart, V., Berioli, M., Rossetto, F.: User scheduling for large multi-beam satellite MIMO systems. In: 2011 Conference Record of the Forty Fifth Asilomar Conference on Signals, Systems and Computers (ASILOMAR), pp. 1800–1804 (2011)
6. Aorpimai, M., Malayavej, V., Navakitkanok, P.: High-fidelity orbit propagator for precise antenna pointing in LEO satellite operation. In: 20th Asia-Pacific Conference on Communication (APCC), pp. 223–226 (2014)
7. Gao, Z.Z., Yang, H.: Arithmetic and analysis of phased array scanning. Chin. Space Sci. Technol. **28**, 60–65 (2008)
8. Hao, W.Y., Pan, D.: Research on optimization of spot-beam satellite antenna pointing to multi-targets. J. Astronaut. **33**, 1788–1793 (2012)
9. Han, X., Guo, X.Z., Wu, J.B.: Research on optimization of spot-beam satellite antenna pointing to multi-targets. Commun. Technol. **48**, 536–540 (2015)
10. Jiang, S.: "Alliance" rocket launching 4 O3b satellites in Kuru. Space Exploration, vol. 2 (2015)
11. Zhu, G.W.: O3b satellite. Satellite Application, vol. 8 (2014)
12. Yin, G.S., Cui, X.H., Dong, H.B., Dong, Y.X., Cui, X.: Quantum-cooperation method for maximum weight perfect matching problem of bipartite graph. J. Comput. Res. Dev. **51**, 2573–2584 (2014)
13. Mohorcic, M., Svigelj, A., Kandus, G., Hu, Y.F., Sheriff, R.E.: Demographically weighted traffic flow models for adaptive routing in packet-switched non-geostationary satellite meshed networks. Comput. Netw. **43**, 113–131 (2003)
14. Nichols, R.A., Moy, D.W., Pattay, R.S.: A fast algorithm for optimal satellite spot beam pointing. IEEE Trans. Aerosp. Electron. Syst. **32**, 1197–1202 (1996)
15. Zhai, Z.A., Tang, C.J.: Fade countermeasure techniques for Ka-band satellite communication links. Chin. Space Sci. Technol. **3**, 55–62 (2010)
16. Lu, J.P.: Analysis of rain attenuation of Ka-band satellite communication system. Radio Commun. Technol. **34**, 6–8 (2008)

End-to-End Stochastic QoS Performance Under Multi-layered Satellite Network

Min Wang[1], Xiaoqiang Di[1,2(✉)], Yuming Jiang[3], Jinqing Li[1], Huilin Jiang[2], and Huamin Yang[1]

[1] School of Computer Science and Technology,
Changchun University of Science and Technology, Changchun, China
dixiaoqiang@cust.edu.cn
[2] Institute of Space Optoelectronics Technology, Changchun, China
[3] Norwegian University of Science and Technology, Trondheim, Norway

Abstract. To meet the growth of real-time and multimedia traffic, the next generation of satellite networks with a guarantee of quality of service (QoS) is indeed, urgent. In this paper, we support the multi-layered satellite network as the scenario, owing to dynamic topology and distinct classification of the generated traffics. We map the satellite network system into a tandem queuing model, which the purpose is devoted to use a mathematical tool for evaluating the performance bounds of per-flow end-to-end networks. For delay-sensitive traffics, we compare two different arrival models–Poisson process and self-similar process. Meanwhile, we apply traditional scheduling strategy to MEO nodes while considering link impairment between a pair of satellites. Finally, we analyze, in a numerically way, which parameters (and how they) influence the per-flow end-to-end performance bounds. Our analysis can be used as a reference to China's future satellite topology and routing algorithms designed and, optimization given a network with performance requirements and constraints.

Keywords: Quality of service · Multi-layered satellite network · Tandem queue model · Transmission path

1 Introduction

With the widespread adoption of multimedia applications in the Internet, how to guarantee quality of service (QoS) for those applications has become a urgent issue in the current research. However, the restriction of geography and technical makes traditional terrestrial wireless network difficult to satisfy users demands. In some particular areas, like islands and isolated mountainous, may not easily deploy adequate infrastructure, even worse in the disasters. Given the advantages of significantly wide coverage and flexible deployment and networking, satellite network plays an essential role in providing better service and worldwide communication environments.

© Springer Nature Singapore Pte Ltd. 2017
Q. Yu (Ed.): SINC 2016, CCIS 688, pp. 182–201, 2017.
DOI: 10.1007/978-981-10-4403-8_17

Recently, according to the altitude of orbits, satellites are classified into Geostationary Earth Orbit (GEO) satellites, Medium Earth Orbit (MEO) satellites and Low Earth Orbit (LEO) satellites. Combining to GEO satellite, LEO and MEO satellite networks have overwhelming advantages, such as small signal attenuation and shorter round trip delays. They are more suitable for real-time service [1]. But there is a case with some delay-sensitive traffics, if the arrival traffics is long distance traffic (LDT) at the LEO layer, because of the increase of hops of the inter-satellite transmission, it will occur congestion easily, higher queueing delay and packet loss [2]. When MEO is regarded as relay node, it will help business flow reducing the exchange of nodes and processing delay. Therefore, the overall network performance, a Multi-Layered Satellite Network (MLSN) has much better than these orbits individually.

Several factors can be found to account for the performance of the MLSN. Apparently, uncertainties and time variances in the spatial environment may be liable to cause various forms of end-to-end performance degradation such as random changes in service properties and space magnetic storms. Under such circumstance, it can generate the impaction of whether the traffics can send successfully. Besides that, the link of a pair of satellites will occur congestion while satellites above hot spot areas are aggregated a huge amount of traffic. Meanwhile, multiple data flows will compete for satellite service resources simultaneously, so that the performance of target flow will be influenced to some extent. Particularly, it is attractive for real-time applications to need a short delay. Thus, on the one hand, it is essential to establish a realistic traffic model for real-time traffics. On the other hand, how to optimize the transmission path and improve the QoS is a pressing problem.

Among existing analysis tools, queueing theory has been proved to be a useful method to deal with traffic model problems in satellite networks. [3] analyzes packet delivery delay in multi-layered satellite networks based on M/M/1 model; [4] employs the stochastic Petri net to evaluate the performance of the double-layered satellite network, but those scheme can lead complex network topologies and heavy traffic load; To avoid traffic congestion, a routing strategy is proposed in [5] to balance the traffic load by using a traffic distribution model. Through these efforts, the satellite network model has been simplified to some extent. Additional research efforts focus on the optimization and design of routing algorithms to meet the QoS requirement for delay and other performances. In the [2,6], a new topological and routing protocol is proposed to evaluate the performance of long-distance traffic; [1] proposes a novel adaptive routing protocol for quality of service (ARPQ) in two-layered satellite network which improves the system performance for delay-sensitive traffic; And in the [7], a unified mathematical framework is proposed to analyze the relationship between network capacity and associated parameters. However, no analytical research considers the end-to-end backlog bound for per-flow and on which parameters (and how they) directly influence the network performances in MLSN.

To better optimize the satellite network performance and improve the data utilization, an effective way to meet the QoS demand is particularly urgent as the current queuing theory can hardly describe the stochastic behaviour of a real satellite network.

Network Calculus (NC) is a new mathematical tool for quantitative performance analysis of computer networks. Based on min-plus algebra and max-plus algebra, this theory is expected to transform a complex network system into an analytically tractable system. The NC concept, originally proposed two decades ago [8,9], has evolved into two branches: Deterministic Network Calculus (DNC) [10] and Stochastic Network Calculus (SNC) [11,12]. Of them, the former mainly focuses on the worst case and cannot address the statistical nature of traffic flow. To tackle this challenge, the latter introduces the probability knowledge into DNC, and thus expands the application of network calculus from deterministic problems to stochastic problems. It can better provide stochastic QoS guarantee to network.

SNC has been widely used to analyze the performance of networks and channels, such as packet switching network [13], LTE network [14], high-speed railway communication network [15], Gilbert-Elliott (G-E) channel [16], Rayleigh fading channel [17], finite Markov-chain channel [18], MIMO wireless channel [19,20], and cognitive radio network [21]. To our knowledge, this is the first work to apply SNC to evaluate the end-to-end performances of MLSN. In this paper, based on the analysis of the MLSN satellite network architecture, the network system is mapped into a tandem queuing system. According to consider the characteristic of spatial traffic flow, we establish a stochastic arrival curve by comparing two different arrival model. Also, we choose the latency-rate scheduling algorithm to provide the MEO nodes services with the impairment of the channel. It aims to build a stochastic service curve. Clearly, the end-to-end performance bounds can be obtained by combining the first two curves with the basic features of SNC. Finally, the numerical analysis shows that this tool can clearly depict the functional relations between these parameters and performance bounds. This paper is just to provide a reference for the design and optimization of topological structure and routing algorithms in the future satellite network.

The rest of this paper is organized as follows. Section 2 maps the multi-layered network system into tandem queue model and presents a description of the channel model. The basic knowledge of SNC is introduced in Sect. 3, including a stochastic arrival curve and stochastic service curve. In Sect. 4 that combine the first two curves with min-plus algebra to derive probabilistic performance bounds, such as end-to-end delay bound and end-to-end backlog bound and so on. Numerical results are discussed in Sect. 5. Finally, Sect. 6 makes a conclusion of the whole paper.

2   System Model

2.1   Multi-layered Satellite Network (MLSN)

In the Fig. 1, the MLSN, including a MEO layer and a LEO layer, is illustrated.

ISL (Inter-Satellite Link): Data can be transmitted at the same layer through ISL. Each satellite is connected with its neighbouring nodes via four ISLs, including intra-orbit links and inter-orbit links. The two satellites in an intra-orbit link are on the same orbit plane, while the two satellites in an inter-orbit link are in different orbits.

ILL (Inter-Layered Link): ILLs are the links connecting the satellites in different layers. The ILL between a LEO satellite and a MEO satellite can be established when the former enters the latter's coverage, otherwise, switched out. Through ILL, different types of data flows can be transferred between LEO satellites and MEO satellites. The link between source node and terminal node must go through one or more end-to-end nodes for data retransmission, whose number, in our view, equals the hop counts [22].

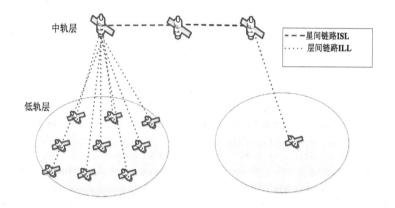

Fig. 1. The multi-layered network system

In this paper, main consideration is given to the satellite network between LEO and MEO layers. The combination of these two layers realizes wireless global coverage, as well as full LEO satellites coverage by MEO satellites. In such a context, we shall guarantee the QoS of this satellite network, whose end-to-end performance shall be achieved through links and reliable connectivity. In order to reduce the computation complexity, there are some assumptions.

(1) Each LEO satellite is only connected to a MEO satellite.
(2) We apply virtual topology strategy in order to shield the dynamics of satellite network topology. A cycle of satellite motion in orbit is divided into several equal time-slicing. The changes of satellite-to-satellite occur only at each point, and the network topology in each time-slicing is same.

Although the MLSN under the current research can provide a lot of network resources, traffic congestion still exists in hot spots [23]. To balance the traffic load, as shown in Fig. 2(a), we assume that one LEO satellite can only be linked to one MEO satellite, and that a group of LEO satellites covered by a MEO satellite can directly transfer the generated data flows to the served MEO, where multiple data flows will meet. Here we take no account of link switching and routing. After receiving the aggregated flows, the MEO satellite will send information to terminal node through one or more hops. Between every pair of MEO satellites, there is a wireless channel.

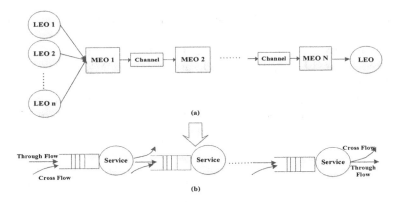

Fig. 2. System model

In the Fig. 2(b), we map the above satellite architecture into a tandem queuing model. The data flows arriving here can be divided into two types: target flow and cross flow. The former is the flow extending from source node to the destination, while the latter is the aggregated flow from other LEO satellite nodes except target flow, and may seriously impact the propagation delay of target flow. The target flow leaving from the source node will pass through multiple nodes before finally arriving at the destination. But beyond that, the arrival and departure of a cross flow can be found in every queue. For the arriving flows, we store them in the corresponding buffer and output them in a first-in-first-out (FIFO) manner. Then a specific scheduling algorithm is provided at the nodes until the flows are successfully transferred to the terminal. We assume that the buffer capacity is infinite so that no overflow exists.

Generally speaking, both the arrival and service processes of an arriving flow are stochastic, so traditional queuing theory can no longer deal with such stochastic performance. Therefore, we apply the SNC in the paper.

2.2 Gilbert-Elliott Channel Model

A satellite channel consists of three main components: transmitting antenna, receiving antenna, and propagation channel. The transmitting antenna sends mass data to the receiving antenna via a wireless satellite channel. The main influencing factor of channel fading is free space loss [24]. In this paper, For the satellite channel, we describe the impairment process as Rayleigh fading channel with a fixed threshold ξ. When the channel is in fading state, the traffic can't be successfully transmitted to the receiver,i.e., $|h(t)| \leq \xi$, otherwise successful at a constant rate, i.e., $|h(t)| > \xi$,where $|h(t)|$ is the envelope process with a Rayleigh probability distribution function and the phase being uniform over $[0, 2\pi)$. To describe the fading channel more correctly, we have mapped it into Gilbert-Elliott (G-E) channel model, as shown in the Fig. 3.

A continuous-time G-E Markov Chain model has two states: ON and OFF, then a peak service rate can be achieved when the model is in good condition.

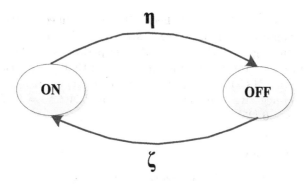

Fig. 3. Gilbert-Elliott channel Model

That is, with the channel ON, the workload is processed at the rate of R_{on}, and the transition rate from OFF to ON is ζ. Likewise, the transition rate from ON to OFF is η. So, the generated matrix Q for GE channel chain can be given as [16],

$$Q = \begin{pmatrix} -\zeta & \zeta \\ \eta & -\eta \end{pmatrix} \tag{1}$$

If the Rayleigh fading channel has an exponential decay rate κ, then, we must have $\eta + \zeta = \kappa$. Through the relationship, we can determine ζ and η as

$$\zeta = \kappa e^{-\xi^2}, \eta = \kappa(1 - e^{-\xi^2}) \tag{2}$$

Consequently the average transmission rate R is:

$$R = \frac{1}{2\theta}(\sqrt{(\zeta - \eta + R_{on}\theta)^2 + 4\zeta\eta} - \zeta - \eta + R_{on}\theta). \tag{3}$$

where $\theta > 0$ is a free parameter.

3   Stochastic Network Calculus

As a basis, SNC contains three processes: the arrival process, service process and departure process in the queuing model, where can be modeled as a traffic model and service model from the first two processes. There are two basic tools including arrival curve and service curve. Arrival curve can be regarded as a constraint to the behaviour of traffic flow, in addition, service curve is an abstract description of service strategy and scheduling strategy, which is exactly provided service lower bound to meet the business QoS requirements, both of curves with min-plus algebra derive probabilistic performance bounds: delay and backlog.

In this paper, we use a series of processes to model an assumed lossless network. Cumulative functions $A(t)$ and $D(t)$ represent the data arrival and departure processes respectively. Of them, $A(s,t)$ is defined as the amount of arriving data accumulated in the period $(s,t]$, while $D(s,t)$ is the amount of arriving data departing from the receiver in the period $(s,t]$. We assume that arrival and service are stationary random processes that are statistically independent. Therefore, $A(s,t) = A(t) - A(s)$ and $D(s,t) = D(t) - D(s)$. $A(t)$ and $D(t)$ are non-negative and increasing on t, and $A(t,t) = 0$, $D(t,t) = 0$, $\forall A(t), D(t) \in \mathcal{F}, t \geq 0$ [25].

3.1   Basic Knowledge

In SNC, there are several different categories in the definition of stochastic arrival curve and stochastic service curve. In this paper, we mainly apply the v.b.c. Stochastic Arrival Curve and Weak Stochastic Service Curve. Definitions are as follows:

Definition 1. *A flow denoted by $A(t) \sim_{vb} < f, \alpha >$ has a v.b.c. stochastic arrival curve $\alpha \in \mathcal{F}$ with a boundary function $f \in \bar{\mathcal{F}}$, if for all $t > 0$ and all $x > 0$ there has*

$$P\{ \sup_{0 \leq s \leq t} \{A(s,t) - \alpha(t-s)\}\} \leq f(x).$$

where $\mathcal{F}$ is non-negative increasing function and $\bar{\mathcal{F}}$ is non-negative non-increasing function.

Definition 2. *A system S denoted by $S \sim_{wb} < g, \beta >$ has a weak stochastic service curve $\beta \in \mathcal{F}$ with a boundary function $g \in \bar{\mathcal{F}}$, if for all $t > 0$ and all $x > 0$ there has*

$$P\{ \sup_{0 \leq s \leq t} \{A \otimes \beta(s) - D(s) > x\}\} \leq f(x).$$

Now we introduce the formula of min-plus convolution:

$$A \otimes B(x) = \inf_{0 \leq x \leq y} [A(x) + B(y-x)].$$

The following basic properties are an essential part within our work, presented in [26, 27].

Theorem 1 (Superposition Property). *Consider the aggregation of N arriving flows with $A_i(t)$, $i = 1, 2, ..., N$. We denote the aggregate arrival process $A_i(t)$, or $A(t) = \sum_{i=1}^{N} A_i(t)$. If $\forall i$, $A_i \sim_{vb} < f_i, \alpha_i >$, then $A(t) \sim_{vb} < f, \alpha >$, where $f(x) = f_1 \otimes \ldots \otimes f_N(x)$, $\alpha(t) = \sum_{i=1}^{N} \alpha_i(t)$.*

Theorem 2 (Concatenation Property). *Regard as a flow through N tandem nodes one by one in a single tandem queue. Each node $i, i = 1, 2, ..., N$ has a stochastic service curve $S^i \sim_{wb} < g^i, \beta^i >$. The overall stochastic service curve obtained by multiple nodes is $S \sim_{wb} < g, \beta >$, where $\beta(t) = \beta^1 \otimes \beta^2 \otimes \ldots \otimes \beta^N(t)$.*

Theorem 3 (Per-flow Service Property). *Suppose the aggregated arriving flow A is composed of two data flows target flow A_1 and cross flow A_2. The system provides the aggregated flow A with a stochastic service curve $S \sim_{wb} < g, \beta >$. If $A_2 \sim_{vb} < f_2, \alpha_2 >$, $\beta_1 \in \mathcal{F}$, then A_1 can obtain the stochastic service curve $S_2 \sim_{wb} < g_1, \beta_1 >$, where $\beta_1 = \beta(t) - \alpha_2(t)$, $g_1(x) = g \otimes f_2(x)$.*

Theorem 4 (Delay Bound). *If for all $t > 0$ and all $x > 0$, the service of network element provides stochastic service curve $S \sim_{wb} < g, \beta >$ and the arrival flow has v.b.c. stochastic arrival curve $A(t) \sim_{vb} < f, \alpha >$. The delay bound $D(t)$ is denoted by*

$$P\{D(t) > h(\alpha + x, \beta)\} \le f \otimes g(x).$$

where $h(\alpha, \beta) = \sup\limits_{s \ge 0} \{inf\{\tau \ge 0, \alpha(s) \le \beta(s + \tau)\}\}$ denotes the maximum horizontal distance between the two function $\alpha(t)$ and $\beta(t)$.

Theorem 5 (Backlog Bound). *If for all $t > 0$ and all $x > 0$, service of network element provides stochastic service curve $S \sim_{wb} < g, \beta >$ and the arrival flow has v.b.c. stochastic arrival curve $A(t) \sim_{vb} < f, \alpha >$, then the backlog bound $B(t)$ is denoted by*

$$P\{B(t) > v(\alpha + x, \beta)\} \le f \otimes g(x)$$

where $v(\alpha, \beta) = \sup\limits_{s \ge 0}\{\alpha(s) - \beta(s)\} \equiv \alpha \oslash \beta(0)$ denotes the maximum vertical distance between the two function $\alpha(t)$ and $\beta(t)$.

In order to better apply SNC to multi-layered satellite network and improve stochastic service guarantees, the next section will introduce a specific stochastic arrival curve and stochastic service curve to model the traffic arrival and node service process.

3.2   Stochastic Arrival Curve

The next generation of satellite network system designed to support multimedia application. This faces a huge challenges to QoS guarantee, as different data flows are expected to meet different performance demands. For example, some real-time applications, such as voice and video, require a low end-to-end delay; and some best-effort traffics, such as email without any specific requirements. Hence it is more and more important to establish different arrival models according to different traffic flows and to meet the QoS demand. In the previous studies, traditional traffic models were Poisson model and ON/OFF model. However, it is fully evidenced in massive current studies that multimedia traffics are characterized by bursty and long range dependence, which can be correctly depicted by self-similarity model [28]. In this paper, we consider the use of two arrival models, traditional Poisson model and self-similarity model, to describe the arrival process in a real-time application, and then compare their influences in the QoS performance analysis.

(1) Poisson Process

The Poisson process, a classical model, is described in the probability distribution of the number of events in per unit interval. $N(t)$, a stable and independent increment, is the number of data arriving at a queuing system during $[0,t]$. The counting process $\{N(t), t \geq 0\}$ can be regarded as a Poisson process. If $N(t) = 0$, the Poisson distribution function will be $P\{N(t) = k\} = e^{(-\lambda t)}(\lambda t)^k/k!$, where λ is the mean arrival rate.

Theorem 6 (Poisson Process). *If the arrival process $A_i(t)$ of aggregated flow is given, the stochastic arrival curve with violation probability ε can be expressed as:*

$$a_n(t) \approx N\lambda t + \sqrt{N\lambda t} \, ln\frac{1}{4\varepsilon}$$

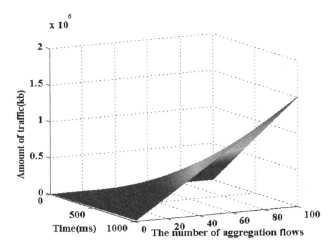

Fig. 4. The relation between the amount of traffic and aggregated flows with the time

As shown in the Fig. 4, we can determine the relations about the amount of total traffic between time interval and aggregation traffic numbers. By making some assumptions such as time $t \in [0, 1000]$ ms, the number of aggregated flows N is from 1 to 100, and some fixed parameters are set, for example, arrival rate $\lambda = 15\,Kb/ms$, and violation probability $\varepsilon = 10^{-3}$. From the Fig. 4, we can observe that, the amount of arriving data will increase sharply if $N \geq 20$, but quite steadily if $N < 20$.

The v.b.c stochastic arrival curve for Poisson process $a_i(t)$ with boundary function $f_i(x)$ is:

$$P\{ \sup_{0 \leq s \leq t} \{A_i(s,t) - a_i(t-s)\}\} \leq a_{f_i}e^{-b_{f_i}x} \tag{4}$$

Here:

$$\begin{cases} \alpha_i(t) = \lambda_i t, \\ f_i(x) = a_{f_i}e^{-b_{f_i}x}. \end{cases}$$

where the average arrival rate $\lambda_i = \frac{\rho_i}{\theta}(e^{\theta \sigma_i} - 1) + \theta_1$, $\alpha_{f_i} = e^{-\theta \theta_1}$ and $b_{f_i} = \theta$. ρ is the packets arrival according to Poisson process with mean rate, and σ is the packet lengths, for $\forall \theta, \theta_1 > 0$.

(2) Self-similarity Process

Fractional Brownian motion (FBM) is widely described as a self-similarity service flow model of aggregated flows with correlated Gaussian increments. It is used to analyze the impact of long range dependence in the network. Self-similarity parameter H is the only similarity standard in the self-similarity process. In case of $H \in (0.5, 1)$, the self-similarity process will be long-range dependent, and the standard form of FBM traffic in the arrival process will be:

$$A(t) = mt + \sqrt{\sigma m} Z_H(t).$$

where the average arrival rate of FBM $m > 0$, σ is the coefficient of dispersion, $Z_H(t)$ is standard FBM process.

The v.b.c stochastic arrival curve for self-similar process $a_i(t)$ with boundary function $f_i(x)$ is:

$$P\{ \sup_{0 \leq s \leq t} \{A_i(s,t) - a_i(t-s)\}\} \leq a_{f_i} e^{-b_{f_i} x^{2(1-H)}}. \tag{5}$$

Here:

$$\begin{cases} \alpha_i(t) = m_i t + \sqrt{\sigma_i m_i} t^H, \\ f_i(x) = a_{f_i} e^{-b_{f_i} x^{2(1-H)}}. \end{cases}$$

where $\alpha_{f_i} = e^{-\theta \theta_1}$ and $b_{f_i} = \theta$, for $\forall \theta, \theta_1 > 0$.

3.3 Stochastic Service Curve

We have developed the Latency-Rate (LR) Server, a universal model, to analyze the traffic scheduling algorithms in the broadband packet network. Some famous scheduling algorithms, such as weighted fair queuing, Virtual Clock, self-clocked fair queuing, weighted round robin, and deficit round robin, are all LR servers. To describe the node service process, we use the LR server model with a week stochastic service curve to describe the feature of node service $\beta(t) = R(t - T)$, where T is the maximum processing latency at the node and R is the minimum service rate (depending upon the channel capacity). From this equation, the relation between for a work-conserving constant rate server counting the effect of packetization can be obtained, that is, $T = L/R$, where L is the maximum packet size, and the boundary function is $g(x) = a_g e^{-b_g x}$ ($a_g = 1$, $b_g = \theta_2$).

According to Theorems 2 and 3, at the node N_i of a MEO satellite, the weak stochastic service curve β^i is available for the target flow A_i aggregated with other cross flows. Thus we can obtain a new stochastic service curve $\beta^i(t)$ and boundary function $g^i(x)$, which are:

$$\beta^i(t) = \beta_i(t) - \alpha_i(t),$$

$$g^i(x) = f_i \otimes g_i(x).$$

The stochastic service curve for MLSN system is:

$$P\{A \otimes \beta^i(t) - D(t) > x\} \leq \alpha_{g_i} e^{-b_{g_i} x}. \tag{6}$$

Here:

$$\begin{cases} \beta^i(t) = R_i(t - T_i), \\ g_i(x) = a_{g_i} e^{-b_{g_i} x}. \end{cases}$$

Now we begin to consider the relationship between the service rate and the fading channel:

The fading channel is governed by Rayleigh fading characteristic [29], as shown below:

$$y(t) = x(t)h(t) + n(t). \tag{7}$$

where $x(t)$ is input signal, $y(t)$ is output signal of satellite system at t, $h(t)$ denotes the channel fading coefficient, $n(t)$ is complex additive white Gaussian noise (AWGN).

In the G-E channel model, the transmission rate R_{on} as the channel is ON can be related to transmission power in the following way:

$$R_{on} = W\log_2(1 + \frac{P\xi^2}{N_o W}). \tag{8}$$

where W is the channel bandwidth, $N_o/2$ is noise power spectral density; P is transmission power; and ξ is a fixed threshold that maps the channel gain $|h(t)|$ into transmission rate.

Then state of G-E channel model depends on channel capacity $C(t)$ and actual transmission rate R_c, which, in turn, depends on AMC (adaptive modulation and coding) in the channel. If $R_c < C(t)$, the channel will be in the ON state, where the transferred data can be received successfully, otherwise, in the OFF state. With the channel ON, the threshold ξ can be expressed as:

$$|h(t)| > \xi = \sqrt{\frac{N_o W}{P}(2^{\frac{R}{W}} - 1)}. \tag{9}$$

The transmission rate R_{on} can be determined by the fixed threshold in (8). For the signal-to-noise ratio γ, there is $\gamma = P/N_o W$. From the above equation, the effect of SNR on service rate R can be seen.

4   Performance Analysis

In the previous chapter, we describe the arrival and service processes of MLSN and obtain a stochastic arrival curve and a stochastic service curve respectively. In this chapter, we will analyze per-flow end-to-end performance parameters.

Corollary 1. *The arriving flow is sent to a MEO node according to a specific scheduling strategy, and then transferred to $N_2, ..., N_n$ before arriving at the destination. This flow is characterized by the stochastic arrival curve $A_i \sim_{vb} < f_i, \alpha_i >$ and the stochastic service curve $S_i \sim_{wb} < g_i, \beta_i >$. So the end-to-end delay and backlog bounds can be expressed as:*

$$P\{D_{end} > h\left(\alpha_1 + x, \beta^{net}\right)\} \leq \underset{1 \leq i \leq n}{\otimes} [f_i \otimes g_i(x)],$$

$$P\{B_{end} > v\left(\alpha_1 + x, \beta^{net}\right)\} \leq \underset{1 \leq i \leq n}{\otimes} [f_i \otimes g_i(x)].$$

where $\beta^{net} = \beta_1 \otimes (\beta_2 - \alpha_2) \otimes \ldots \otimes (\beta_N - \alpha_N)(t)$.

Proof. Now let's derive from the boundary function in the Poisson arrival process with the help of min-plus algebra. At first, we consider two nodes:

$$\beta_1 \otimes (\beta_2 - \alpha_2)(t) = R_1(t - T_1) \otimes [R_2(t - T_2) - \lambda_2 t]$$
$$= \underset{0 \leq s \leq t}{inf} [(R_1 - R_2 + \lambda_2)s + (R_2 - \lambda_2)t - (R_1 T_1 + R_2 T_2)].$$

It is clear that, only with the theory of $R_1 + \lambda_2 \geq R_2$, sufficient service flows can be guaranteed and the best results may be yielded. We use $s = 0$ to obtain the minimum value:

$$\beta_1 \otimes (\beta_2 - \alpha_2)(t) \otimes (\beta_3 - \alpha_3)(t) = \underset{0 \leq s \leq t}{inf} [(R_2 - \lambda_2 - R_3$$
$$+ \lambda_3)s + (R_3 - \lambda_3)t - \sum_{i=1}^{3} R_i T_i].$$

Next, we consider three nodes:

$$\beta_1 \otimes (\beta_2 - \alpha_2)(t) \otimes (\beta_3 - \alpha_3)(t) = \underset{0 \leq s \leq t}{inf} [(R_2 - \lambda_2 - R_3$$
$$+ \lambda_3)s + (R_3 - \lambda_3)t - \sum_{i=1}^{3} R_i T_i].$$

When $(R_2 - \lambda_2)t > (R_3 - \lambda_3)t$, we set $s = 0$, thus:

$$\beta_1 \otimes (\beta_2 - \alpha_2)(t) \otimes (\beta_3 - \alpha_3)(t) = (R_3 - \lambda_3)t - \sum_{i=1}^{3} R_i T_i.$$

When $(R_2 - \lambda_2)t \leq (R_3 - \lambda_3)t$, we set $s = 0$, thus:

$$\beta_1 \otimes (\beta_2 - \alpha_2)(t) \otimes (\beta_3 - \alpha_3)(t) = (R_2 - \lambda_2)t - \sum_{i=1}^{3} R_i T_i.$$

In this way, we can obtain the stochastic service curve of the whole network:

$$\beta^{net} = \underset{2 \leq i \leq N}{inf} [(R_i - \lambda_i)t] - \sum_{i=1}^{N} R_i T_i. \tag{10}$$

Similarly, we can obtain the stochastic service curve in the self-similarity arrival process:

$$\beta^{net} = \inf_{2 \leq i \leq N} [(R_i - m_i)t - \sqrt{\sigma_i m_i}t^{H_i}] - \sum_{i=1}^{N} R_i T_i. \tag{11}$$

With (10) or (11), apply β^{net} to Corollary 1, the maximum horizontal distance can be readily obtained. When the arrival model is Poisson process:

$$d_{end} = \frac{NRT}{R-\lambda} - \frac{N(\frac{1}{b_f} + \frac{1}{b_g})}{R-\lambda} \times \ln \frac{\varepsilon}{Nab(\frac{1}{b_f}+\frac{1}{b_g})}. \tag{12}$$

$$b_{end} = \lambda t + NRT - N(\frac{1}{b_f} + \frac{1}{b_g}) \times \ln\frac{\varepsilon}{Nab(\frac{1}{b_f} + \frac{1}{b_g})}. \tag{13}$$

When the arrival model is self-similarity process:

$$d_{end} \approx \frac{[-\frac{\ln \varepsilon}{\theta} - \theta_1]^{\frac{1}{2(1-H)}} + NRT}{R - m - \sqrt{\sigma m}}. \tag{14}$$

$$b_{end} \approx mt + \sqrt{\sigma m}t^H + NRT + [-\frac{\ln \varepsilon}{\theta} - \theta_1]^{\frac{1}{2(1-H)}}. \tag{15}$$

Proof. To analyze the boundary functions of the whole network, we adopt the following lemma:

Lemma 1. For any positives a_k, b_k, $(k = 1, ..., K)$ and $x > 0$, the following equation is:

$$\inf_{x_1 + \cdots x_k = K} \sum_{k=1}^{K} a_k e^{-b_k x_k} = e^{-\frac{x}{w}} \prod_{k=1}^{k} (a_k b_k w)^{\frac{1}{b_k w}}$$

where $w = \sum_{k=1}^{K} \frac{1}{b_k}$, the related content is given in [27].

With the lemma above, the boundary function can be expressed as:

$$\bigotimes_{1 \leq i \leq N} \{f_i \otimes g_i(x)\} = \bigotimes_{1 \leq i \leq N} \inf_{0 \leq y \leq x} \{\alpha_{f_i} e^{-b_{f_i} y} + \alpha_{g_i} e^{-b_{g_i}(x-y)}\}$$

$$= \bigotimes_{1 \leq i \leq N} \alpha_i e^{-b_i x}$$

$$= e^{-\frac{x}{w}} \prod_{i=1}^{N} (a_i b_i w)^{\frac{1}{b_i w}}$$

where $a_i = (\alpha_{f_i} b_{f_i} w_i)^{\frac{1}{b_{f_i} w_i}} \times (\alpha_{g_i} b_{g_i} w_i)^{\frac{1}{b_{g_i} w_i}}, \, b_i = \frac{1}{w_i}, \, w_i = \frac{1}{b_{f_i}} + \frac{1}{b_{g_i}}, \, w = \sum\limits_{i=1}^{N} \frac{1}{b_i} =$

$\sum\limits_{i=1}^{N} (\frac{1}{b_{f_i}} + \frac{1}{b_{g_i}}).$

Finally, we can derive the end-to-end delay and backlog performance bounds from SNC.

If the arrival model is Poisson distribution, then:

$$\left(P\{D_{end} > h(\alpha_1 + x, \inf_{2 \leq i \leq N} [(R_i - \lambda_i)t] - \sum_{i=1}^{N} R_i T_i)\} \right) \leq e^{-\frac{x}{w}} \prod_{i=1}^{N} (a_i b_i w)^{\frac{1}{b_i w}}$$

(16)

$$\left(P\{B_{end} > v(\alpha_1 + x, \inf_{2 \leq i \leq N} [(R_i - \lambda_i)t] - \sum_{i=1}^{N} R_i T_i)\} \right) \leq e^{-\frac{x}{w}} \prod_{i=1}^{N} (a_i b_i w)^{\frac{1}{b_i w}}$$

(17)

where $a_1 = \lambda_1 t, \, w = \sum\limits_{i=1}^{N} \frac{1}{b_{g_i}}.$

If the arrival model is self-similarity, then:

$$P\{D_{end} > h(\alpha_1 + x, \inf_{2 \leq i \leq N} [(R_i - m_i)t - \sqrt{\sigma_i m_i} t^{H_i}] - \sum_{i=1}^{N} R_i T_i)\}$$

(18)

$$\leq (\bigotimes_{1 \leq i \leq N} a_{f_i} e^{-b_{f_i} x^{2(1-H)}}) \otimes (e^{-\frac{x}{w}} \prod_{i=1}^{N} (a_{g_i} b_{g_i})^{\frac{1}{b_{g_i} w}}).$$

$$P\{B_{end} > v(\alpha_1 + x, \inf_{2 \leq i \leq N} [(R_i - m_i)t - \sqrt{\sigma_i m_i} t^{H_i}] - \sum_{i=1}^{N} R_i T_i)\}$$

(19)

$$\leq (\bigotimes_{1 \leq i \leq N} a_{f_i} e^{-b_{f_i} x^{2(1-H)}}) \otimes (e^{-\frac{x}{w}} \prod_{i=1}^{N} (a_{g_i} b_{g_i})^{\frac{1}{b_{g_i} w}}).$$

where $\alpha_1 = m_1 t + \sqrt{\sigma_1 m_1} t^{H_1}, \, w = \sum\limits_{i=1}^{N} \frac{1}{b_{g_i}}.$

Let the right side of (16), (17), (18) and (19) equal to violation probability ε, then can get the end-to-end delay and backlog bounds.

5 Numerical Analysis

In this chapter, we derive the numerical results of per-flow end-to-end delay and backlog bounds in the MLSN, while contrasting two different types of arrival model: Poisson process and self-similar process. The consequences are achieved based on the previous sections, after discussion. Finally, we draw a conclusion that can provide insights into the influence of different parameters on QoS performances in the MLSN.

We use MATLAB simulation tool to implement analytical results. During the simulation, we set the following fixed parameters at first: number of aggregated flows is 20; maximum packet size is 10^4 bytes; exponential decay rate $\kappa = 100$; and θ, θ_1 and θ_2, which can make the performance bounds as tight as possible after being adjusted.

From the Fig. 5, we can visually find that, in different traffic models, the delay performance will become worse as the number of nodes increases. That is, the more hops, the greater influence on performance bounds. We also examine the effect of delay bound and violation probability. Violation probability is defined as the probability that the packet delay exceeds the stochastic delay bound. Once the violation probability varies, the delay will increase. When the traffic model is Poisson or self-similarity process, the results are shown in Fig. 5(a) or (b) respectively. It can be seen that, under the constraints of the same parameters, the stochastic delay of Poisson model is smaller than that of self-similarity traffic model. It is demonstrable that, for real-time communications, the well-known Poisson model is more efficient than self-similarity. In the real network, we can choose an efficient path for a certain topological structure to achieve the best performance.

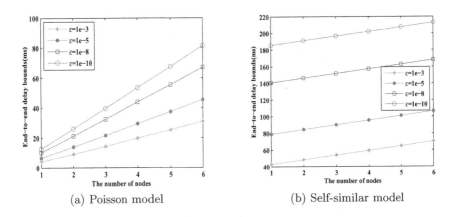

(a) Poisson model   (b) Self-similar model

Fig. 5. End-to-end delay bounds vary from the number of nodes and different violation probability under SNR as $\gamma = 25\,\mathrm{dB}$ and the arrival rate in different models $m = \rho = 15\,\mathrm{Kb/ms}$

As shown in the Fig. 6, in the self-similarity model, the end-to-end delay bound increases with H parameter under the condition of the self-similar model. When H rises to 0.9, the delay bound will increase rapidly. It indicates that the self-similar traffic model can reduce delay.

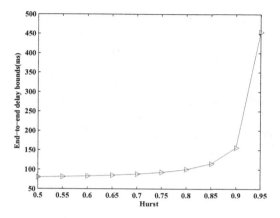

Fig. 6. End-to-end delay bounds with the Hurst values under SNR as $\gamma = 25\,\text{dB}$ and the arrival rate of self-similar process $m = 15\,\text{Kb/ms}$

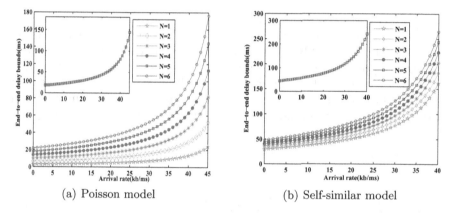

(a) Poisson model   (b) Self-similar model

Fig. 7. End-to-end delay bounds vary from the number of nodes and different violation probability under SNR as $\gamma = 25\,\text{dB}$ and the arrival rate in different models $m = \rho = 15\,\text{Kb/ms}$

We first discuss the sub-graph in Fig. 7. If the traffic model is a Poisson process, the horizontal axis will represent the arrival rate from 0 to 45 Kb/ms and the vertical axis will indicate the end-to-end delay bound. Obviously, the delay will increase rapidly when the arrival rate goes up to 40 Kb/ms, otherwise level off. It can be seen from the Fig. 7(a) that, in the future network design, the maximum arrival rate cannot exceed 40, or else excessive delay will be resulted in to degrade the performance. Likewise, from the Fig. 7(b), we can also observe that the change in arrival rate influences the performance. The comparison between the two pictures shows that, the end-to-end delay bound in the Poisson arrival process is lower than that in the self-similar process.

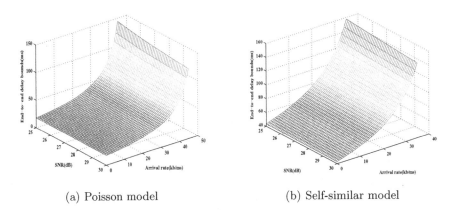

(a) Poisson model   (b) Self-similar model

Fig. 8. End-to-End delay bounds vary from different SNR and arrival rate under the number of nodes $N = 5$ and the violation probability $\varepsilon = 10^{-3}$

In the Fig. 8, the 3D relationships among SNR, arrival rate and end-to-end delay bound are illustrated. We select several points from the picture for comparison. Obviously, as the SNR increases, the end-to-end delay bound will decrease with arrival rate. This is because that, with the increase of SNR, the satellite channel is more likely to be in better condition. The comparison between Fig. 8(a) and (b) shows that, an optimum delay bound can be obtained through the trade off between SNR and arrival rate under the Poisson constraint.

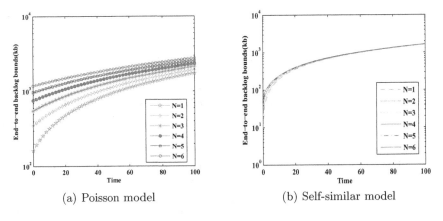

(a) Poisson model   (b) Self-similar model

Fig. 9. End-to-End backlog bounds vary from time and node number under SNR as $\gamma = 25$ dB and the violation probability $\varepsilon = 10^{-3}$

In the Fig. 9, the impact of the number of nodes on end-to-end backlog bound is illustrated. At first, we set some fixed parameters: average arrival rate in Fig. 9(a), $\rho = 15$ Kb/ms; and $m = 15$ Kb/ms in Fig. 9(b); and Hurst parameter $H = 0.5$. Obviously, the end-to-end backlog bound increases with the number of nodes. The comparison between Fig. 9(a) and (b) shows that, the impact of Poisson arrival process in Fig. 9(a) on backlog is worse than that in Fig. 9(b). This is because that, in the self-similar arrival process, the traffic is characterized by bursty and instability, and the nodes need more services to guarantee QoS. The final conclusion is that better backlog can be brought by choosing the self-similar arrival model and optimizing the number of nodes.

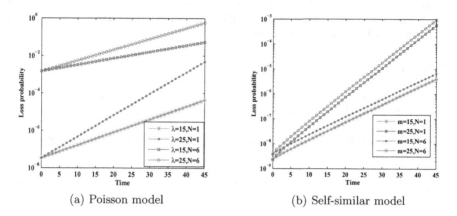

(a) Poisson model   (b) Self-similar model

Fig. 10. Loss probability vary from different arrival rate and node number under SNR as $\gamma = 25\,\text{dB}$

We can see from the Fig. 10 that, the packet loss probability increases with time. After changing some parameters, for example, changing the number of nodes N from 1 to 6 or the arrival rate (15 Kb/ms, 25 Kb/ms), the packet loss probability will be $P(B^{end}(t) > b_{max})$, where the size of fixed buffer (bmax) is $b_{max}) = 2000\,\text{Kb}$. The Fig. 10 shows that, if a given probability is reached, the number of nodes shall be optimized in addition to the reduction of arrival rate. Also, by considering both Figs. 9 and 10, we learn that the success rate of data transmission can be significantly improved when the arrival model is Poisson model.

6   Conclusion

The main contribution of this paper is to use SNC to analyze the per-flow end-to-end performance bounds in the MLSN. In this paper, multiple data flows are aggregated and then arrive at a MEO satellite. The arrival process is simulated in two models: Poisson model and self-similar model. In addition, appropriate LR scheduling strategies are provided at the nodes. In the analysis, we map a complex network structure into a tandem queuing system, which, with SNC, is used to derive the functional relations between end-to-end delay/backlog bounds and other parameters. Finally, through numerical simulation, we can see more clearly that some parameters such as the violation probability, the number of nodes, SNR and arrival rate, have an impact on these performance bounds. The future improvement in this regard is to bring in simulation and numerical analysis for comparison, in order to better verify the applicability of SNC in the MLSN.

Acknowledgements. This work is supported by the National Science Foundation of China (863 Project) under granted No. 2015AA015701, the project of Provincial Science and Technology Department of Jilin (No. 20140414070GH) and Science and Technology of Changchun under granted No. 14GH001.

References

1. Bayhan, S., Gür, G., Alagz, F.: Performance of delay-sensitive traffic in multi-layered satellite IP networks with on-board processing capability. Int. J. Commun. Syst. **20**, 1367–1389 (2007)
2. Lee, J., Kang, S.: Satellite over satellite (SOS) network: a novel architecture for satellite network. In: Proceedings of Nineteenth Annual Joint Conference of the IEEE Computer and Communications Societies in INFOCOM, pp. 392–399 (2000)
3. Kawamoto, Y., et al.: Assessing packet delivery delay in multi-layered satellite networks. In: 2012 IEEE International Conference on in Communications (ICC), pp. 3311–3315 (2012)
4. Wu, F.G., et al.: Performance evaluation on a double-layered satellite network. Int. J. Satell. Commun. Netw. **23**, 359–371 (2005)
5. Soret, B., Aguayo-Torres, M.C., Tomás, J.: Maximum delay-constrained source rate over a wireless channel. In: Proceedings of the 2nd International Conference on Performance Evaluation Methodolgies and Tools, Entrambasaguas, Nantes, France, PP. 1–9 (2007)
6. Zhou, Y., Sun, F., Zhang, B.: A novel QoS routing protocol for LEO and MEO satellite networks. Int. J. Satell. Commun. Netw. **25**, 603–617 (2007)
7. Runzi, L., et al.: Capacity analysis of two-layered LEO/MEO satellite networks. In: Vehicular Technology Conference (VTC Spring) (2015)
8. Cruz, R.L.: A calculus for network delay, part I: network elements in isolation. IEEE Trans. Inf. Theor. **37**, 114–131 (1991)
9. Cruz, R.L.: A calculus for network delay, part II: network analysis. IEEE Trans. Inf. Theor. **37**, 132–141 (1991)
10. Le Boudec, J.Y., Thiran, P.: Network Calculus Parts II and III - A Theory of Deterministic Queuing Systems for the Internet. LNCS. Springer, Heidelberg (2004)
11. Fidler, M., Rizk, A.: A guide to the stochastic network calculus. IEEE Commun. Surv. Tutor. **17**, 92–105 (2015)
12. Jiang, Y.: Stochastic network calculus for performance analysis of Internet networks An overview and outlook. In: IEEE International Conference on Computing, Networking and Communications, pp. 638–644 (2012)
13. Geyer, F.: End-to-end flow-level quality-of-service guarantees for switched networks. Network Architectures and Services (2015)
14. Xin, C., Lei, Z., Xiang, X., Wan, D.: A stochastic network calculus approach for the end-to-end delay analysis of LTE networks. Chin. J. Comput. **35**, 30–35 (2011)
15. Li, Y., Lei, L., Zhong, Z., et al.: Performance analysis for high-speed railway communication network using stochastic network calculus. In: IET International Conference on Wireless, Mobile and Multimedia Networks, pp. 100–105 (2013)

16. Fidler, M.: WLC15-2: a network calculus approach to probabilistic quality of service analysis of fading channels. In: GLOBECOM, pp. 1–6 (2006)
17. She, H., Lu, Z., Jantsch, A., et al.: Modeling and analysis of Rayleigh fading channels using stochastic network calculus. In: Wireless Communications and Networking Conference (WCNC), pp. 1056–1061. IEEE (2011)
18. Lin, S., et al.: Finite state Markov modelling for high speed railway wireless communication channel. In: IEEE in Global Communications Conference (GLOBECOM), pp. 5421–5426 (2012)
19. Mahmood, K., Rizk, A., Jiang, Y.: On the flow-level delay of a spatial multiplexing MIMO wireless channel. In: IEEE International Conference on Communications (ICC), pp. 1-6 (2011)
20. Mahmood, K., Vehkapera, M., Jiang, Y.: Delay constrained throughput analysis of a correlated MIMO wireless channel. In: IEEE International Conference on Computer Communications and Networks, pp. 1–7 (2011)
21. Gao, Y., Jiang, Y.: Performance analysis of a cognitive radio network with imperfect spectrum sensing. In: INFOCOM IEEE Conference on Computer Communications Workshops, pp. 1-6 (2010)
22. Chen, C., Ekici, E.: A routing protocol for hierarchical LEO/MEO satellite IP networks. Wirel. Netw. **11**, 507–521 (2005)
23. Mohorcic, M., et al.: Demographically weighted traffic flow models for adaptive routing in packet-switched non-geostationary satellite meshed networks. Comput. Netw. **43**, 113–131 (2003)
24. Patnaik, B., Sahu, P.K.: Inter-satellite optical wireless communication system design and simulation. IET Commun. **6**, 2561–2567 (2012)
25. Fidler, M.: An end-to-end probabilistic network calculus with moment generating functions. In: 14th IEEE International Workshop on Quality of Service (2005)
26. Jiang, Y.: A basic stochastic network calculus. ACM SIGCOMM Comput. Commun. Rev. **36**, 123–134 (2006)
27. Jiang, Y., Liu, Y.: Stochastic Network Calculus, vol. 1. Springer, Heidelberg (2010)
28. Na, Z., et al.: Research on aggregation and propagation of self-similar traffic in satellite network. Int. J. Hybrid Inf. Technol. **8**, 325–338 (2015)
29. Mahmood, K., Vehkaper, M., Jiang, Y.: Delay constrained throughput analysis of SISO. In: 2012 3rd IEEE International Conference on Network Infrastructure and Digital Content (IC-NIDC) (2012)

A Non-stationary 3-D Multi-cylinder Model for HAP-MIMO Communication Systems

Zhuxian Lian$^{(\boxtimes)}$, Lingge Jiang$^{(\boxtimes)}$, and Chen He

Department of Electronic Engineering,
Shanghai Jiao Tong University, Shanghai 200240, China
{lianzhuxian,lgjiang}@sjtu.edu.cn

Abstract. Due to the movement of the receiver, the time-variant transfer function of the radio channel of the high altitude platform (HAP) is a non-stationary process. A theoretical non-stationary three dimensional (3-D) multi-cylinder HAP multiple-input multiple-output (MIMO) channel model is proposed in this paper. The space-time correlation function of the proposed 3-D HAP-MIMO channel is proposed in this paper. In addition, we propose a corresponding simulation model. Numerical results show that the simulation model could fit to the proposed 3-D HAP-MIMO channel model very well. This paper also investigates the capacity of spatially and temporally correlated HAP-MIMO channel achieved with uniform linear arrays (ULAs).

1 Introduction

As the radio spectrum is a limited resource and future generation multimedia applications demand for increasingly capacity, the high altitude platform (HAP) is considered as a promising technology for next generation wireless communication [1–3]. The HAP is an airship or aircraft operating in lower stratosphere at an altitude of 17–22 km. It has advantages both satellite communication system (SCS) and terrestrial communication system (TCS) [4], and it has attracted considerable attention worldwide [5].

Multiple-input multiple-output (MIMO) technology has the ability to significantly enhance the performance of wireless communication systems in rich multipath environments, without increasing transmit power and bandwidth [6]. As a new emerging technology, the challenge is to investigate the application of MIMO technology to HAP communication systems. The impairments of the signal of HAP-MIMO communication systems are mainly caused by the environment around the terrestrial mobile station (TMS), and the establishment of a particular geometry model is highly critical [7]. In [8,9], three-dimensional (3-D) cylinder model are proposed, and the closed-form joint space-time correlation function is derived for 3-D nonisotropic scattering environment. A geometry-based single-bounce (GBSB) HAP model is proposed in [10], and it assumes all the scatterers are located in an ellipsoid with transmitter and receiver as foci. However, the influence of the HAP elevation angle is not considered for the performance evaluation. In [11], a statistical model for satellite/HAP-MIMO

© Springer Nature Singapore Pte Ltd. 2017
Q. Yu (Ed.): SINC 2016, CCIS 688, pp. 202–214, 2017.
DOI: 10.1007/978-981-10-4403-8_18

channel is proposed, and specific distributions for scatterers are used. In [7], a 3-D GBSB model for HAP-MIMO channels is proposed, and it assumes that the scatterers which are located in the vicinity of the mobile terminal are nonuniformly distributed within a cylinder. The model considers that the distances between scatterers and mobile station are subject to a hyperbolic distribution, but they are only applicable to the case that the azimuth angles of departure from scatterers are uniformly distributed [12].

Due to the movement of the receiver, the time-variant transfer function of the radio channel of the high altitude platform (HAP) is a non-stationary process [13]. A non-stationary channel model based on scattering volumes for satellite communication systems is proposed in [14]. According to [13,14], the models in [7,10,11] could not describe the non-stationary properties of HAP-MIMO channel. A non-stationary 3-D multi-cylinder channel model is proposed, and the space-time correlation function is also investigated in this paper. A corresponding simulation model is also proposed in this paper. In addition, this paper also investigates the capacity of spatially and temporally correlated HAP-MIMO channel achieved with uniform linear arrays (ULAs).

2   3-D HAP-MIMO Channel Model

In this section, we introduce a 3-D reference model for HAP-MIMO channel, as shown in Figs. 1 and 2. We consider a fixed stratospheric base station (SBS) and a moving receiver, deployed with N_T transmit and N_R receive antenna elements, respectively. The distances between antenna elements are δ_T at transmitter and δ_R at receiver. Let $p\,(p = 1, 2, \cdots, N_T)$ and $l\,(l = 1, 2, \cdots, N_R)$ denotes p-th transmit and l-th antenna element, respectively. Let us assume that SBS is free of local scatterers and the receiver is located in the bottom center of vertical cylinder which is used to mimic the scattering surfaces [8,9]. Based on the tapped delay line (TDL), i-th $(i = 1, 2, \cdots, I)$ vertical cylinder denotes i-th tap, and I is the total number of taps. There are N_i omnidirectional scatterers on the surface of the i-th cylinder. Then, the n_i-th scatterer is denoted by S_{n_i}. The parameters of the proposed model are summarized in Table 1.

Based on the 3-D HAP-MIMO channel model, the channel impulse response (CIR) between the p-th transmit and l-th receive antenna element can be expressed as

$$h_{pl}\,(t, \tau) = \sum_{i=1}^{I} h_{i,pl}\,(t)\,\delta\,(\tau - \tau_i) \tag{1}$$

where $h_{i,pl}\,(t)$ and τ_i denote the complex time-variant tap coefficient and the discrete propagation delay of the i-th tap.

The complex tap coefficient of sub-channel $p - l$ $h_{i,pl}\,(t)$ is a superposition of the line-of-sight (LoS) components $h_{i,pl}^{\mathrm{LoS}}\,(t)$ and non-line-of-sight (NLoS) components $h_{i,pl}^{\mathrm{NLoS}}\,(t)$, and it can be expressed as

$$h_{i,pl}\,(t) = \delta\,(i - 1)\,h_{i,pl}^{\mathrm{LoS}}\,(t) + h_{i,pl}^{\mathrm{NLoS}}\,(t) \tag{2}$$

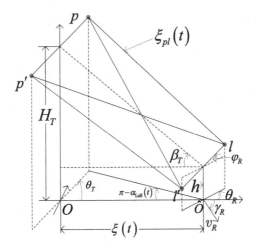

Fig. 1. LoS paths of the 3-D HAP-MIMO channel model.

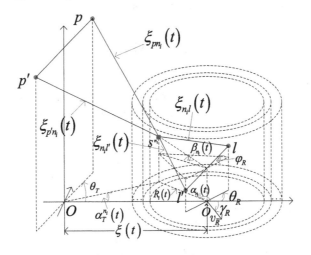

Fig. 2. NLoS paths of the 3-D HAP-MIMO channel model.

where

$$h_{1,pl}^{\text{LoS}}(t) = \sqrt{\frac{K}{K+1}} \exp\left\{-j2\pi\xi_{pl}(t)/\lambda + j2\pi f_{\max}\cos\left(\alpha_{\text{LoS}}(t) - \gamma_R\right)t\right\} \quad (3)$$

$$h_{i,pl}^{\text{NLoS}}(t) = \sqrt{\frac{\Omega_i}{\delta(i-1)K+1}} \lim_{N_i \to \infty} \frac{1}{\sqrt{N_i}} \times \sum_{n_i=1}^{N_i} g_{n_i} \times \exp\left(j\varphi_{n_i}\right)$$

$$\times \exp\left\{-j2\pi/\lambda \times \left(\xi_{pn_i}(t) + \xi_{n_il}(t)\right) + j2\pi f_{\max}\cos\left(\alpha_{n_i}(t) - \gamma_R\right)\cos\beta_{n_i}(t)t\right\}$$

$$(4)$$

Table 1. Definition of the parameters used in the proposed model

$\xi(t)$	Distance between the center O of projection of the SBS antenna elements to the xy-plane and the lower centre $\hat{O}$ of the cylinder
β_T	Elevation angle of the SBS relative to receiver
$\alpha_{\mathrm{LoS}}(t)$	Azimuth angle of arrival (AAOA) of line-of-sight (LOS) paths of the receiver
H_T, h	Height of SBS and receiver, respectively
$R_i(t)$	Radius of the i-th cylinder
θ_T, θ_R	Orientation of antenna elements of the SBS and receiver in the xy-plane relative to x-axis, respectively
φ_R	Elevation angle of antenna elements of the receiver relative to xy-plane
$\alpha_T^{n_i}(t)$	Azimuth angle of arrival (AAOA) at the n_ith scatterer
$\alpha_{n_i}(t)$	Azimuth angle of departure (AAOD) from the n_ith scatterer
$\beta_{n_i}(t)$	Elevation angle of departure (EAOD) from the n_ith scatterer
γ_R, v_R	Moving directions and velocity of the receiver
β_n	The elevation angle of departure (EAOD) from the nth scatterer.
$\xi_{pl}(t)$	Distance between the p-th SBS antenna element and the l-th receive antenna element
$\xi_{pn_i}(t), \xi_{p'n_i}(t)$	Distance between the p-th, p'-th SBS antenna elements and the n_i-th scatterer, respectively
$\xi_{n_i l}(t), \xi_{n_i l'}(t)$	Distance between the n_ith scatterer and l-th, l'-th receive antenna elements, respectively

where $f_{\max} = v_R/\lambda$ is the receiver maximum Doppler frequency, λ is the carrier wavelength. K and Ω_i denote Rician factor and transmitted power for the i-th tap. g_{n_i} represents the amplitude of the n_i-th scattered wave such that $N_i^{-1} \sum_{n_i=1}^{N_i} E\left[|g_{n_i}|^2\right] = 1$ as $N_i \to \infty$. It is assumed that the phase φ_{n_i} is random variable uniformly distributed in the interval $[-\pi, \pi)$ and independent from the angle of the departure and the angle of arrival.

Using the laws of sines and cosines and the approximate relation $\sqrt{1+x} \approx 1 + x/2$, $\sin x \approx x$, and $\cos x \approx 1$ for a small x, we can derive

$$\xi_{pl}(t) \approx (\xi(t) - \xi_{t_1} + \xi_{r_1})/\cos\beta_T \tag{5}$$

$$\xi_{pn_i}(t) \approx (\xi(t) - \xi_{t_1} - \xi_{t_2}\sin\alpha_{n_i}(t)R_i(t)/\xi(t))/\cos\beta_T \tag{6}$$

$$\xi_{n_i l}(t) \approx R_i(t)/\cos\beta_{n_i}(t) - \xi_{r_1}\cos\alpha_{n_i}(t)\cos\beta_{n_i}(t) - \xi_{r_2}\sin\alpha_{n_i}(t)\cos\beta_{n_i}(t)$$
$$- \xi_{r_3}\sin\beta_{n_i}(t) \tag{7}$$

where

$$\xi_{t_1} = 1/2 \cdot (N_T + 1 - 2p)\delta_T \cos\theta_T \tag{8}$$

$$\xi_{t_2} \approx 1/2 \cdot (N_T + 1 - 2p) \, \delta_T \sin \theta_T \tag{9}$$

$$\xi_{r_1} \approx 1/2 \cdot (N_R + 1 - 2l) \, \delta_R \cos \theta_R \cos \varphi_R \tag{10}$$

$$\xi_{r_2} \approx 1/2 \cdot (N_R + 1 - 2l) \, \delta_R \sin \theta_R \cos \varphi_R \tag{11}$$

$$\xi_{r_3} \approx 1/2 \cdot (N_R + 1 - 2l) \, \delta_R \sin \varphi_R \tag{12}$$

As the number of scatterers tends to infinity, the discrete angles $\alpha_{n_i}(t)$ and $\beta_{n_i}(t)$ can be replaced by continuous $\alpha_i(t)$ and $\beta_i(t)$, respectively. We use von Mises probability density function (PDF) which is defined as [15]

$$f(\alpha_i(t)) = \frac{\exp(\kappa \cos(\alpha_i(t) - \mu_i(t)))}{2\pi I_0(\kappa)} \tag{13}$$

where $I_0(\cdot)$ is the zeroth-order modified Bessel function of the first kind, $\mu_i(t) \in [-\pi, \pi]$ is the mean AAOD, and κ controls the spread around the mean. The Parsons PDF is defined as [16]

$$f(\beta_i(t)) = \frac{\pi \cos(\pi \beta_i(t)/(2\beta_{\max}))}{|4\beta_{\max}|} \tag{14}$$

where the parameter $\beta_{\max}$ is the absolute value of scatterers maximum elevation angle.

3   Time-Varying Parameters

The non-stationary property of HAP-MIMO channel is mainly caused by the movement of receiver and scatterer. In order to describe the non-stationary property, the geometrical relationships among the receiver and scatterer should be updated. Figure 3 shows the projection of a 3-D HAP-MIMO channel model.

The time-variant distance $\xi(t)$ can be expressed as

$$\xi(t) = \sqrt{(\xi(t_0))^2 + (v_R(t - t_0))^2 - 2\xi(t_0) v_R(t - t_0) \cos(\pi - \gamma_R)} \tag{15}$$

where $\xi(t_0)$ denotes the initial distance at time $t = t_0$.

The AAoA of LoS components $\alpha_{\mathrm{LoS}}(t)$ can be expressed as

$$\alpha_{\mathrm{LoS}}(t) = \alpha_{\mathrm{LoS}}(t_0) - \arccos\left(\frac{\xi(t_0) + v_R(t - t_0) \cos \gamma_R}{\xi(t)}\right) \tag{16}$$

where $\alpha_{\mathrm{LoS}}(t_0) \approx \pi$.

Auxiliary variable θ_1 can be expressed as

$$\theta_1 = \arcsin\left(\frac{v_R(t - t_0)}{\xi(t)} \sin(\pi - \gamma_R)\right) \tag{17}$$

At the time t, $R_i(t)$ can be expressed as

$$R_i(t) = \sqrt{(R_i(t_0))^2 + (v_R(t - t_0))^2 - 2R_i(t_0) v_R(t - t_0) \cos(\mu(t_0) - \gamma_R)} \tag{18}$$

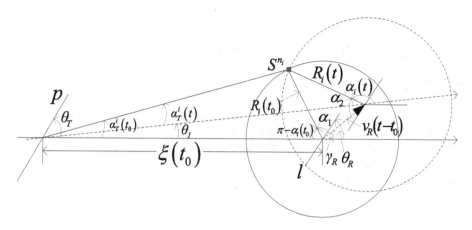

Fig. 3. NLoS paths of the 3-D HAP-MIMO channel model.

At the time t, $\mu(t)$ can be expressed as

$$\mu(t) = \pi + \gamma_R - \theta_1 - \alpha_2 \tag{19}$$

where $\alpha_2 = \arcsin\left(R_i(t_0)/R_i(t)\sin(\mu(t_0) - \gamma_R)\right)$.

4   Space-Time Correlation Function

In this section, the space-time correlation function for HAP-MIMO channel is derived. The normalized space-time correlation function between $p - l$ link and $p' - l'$ link is defined as

$$R_{i,pl,p'l'}(\delta_T, \delta_R, t, \Delta t) = \frac{E\left[h_{i,pl}(t)\, h^*_{i,p'l'}(t + \Delta t)\right]}{\sqrt{\Omega_{i,pl}\Omega_{i,p'l'}}} \tag{20}$$

where $(\cdot)^*$ and $E[\cdot]$ denote complex conjugate operation and statistical expectation operation, $p, p' \in \{1, 2, \cdots, N_T\}$, $l, l' \in \{1, 2, \cdots, N_R\}$. —In the case of LoS components

$$R^{\text{LoS}}_{i,pl,p'l'}(\delta_T, \delta_R, t, \Delta t) = \frac{\delta(i - 1)K}{K + 1} e^{-j\frac{2\pi}{\lambda}\left(\xi_{pl}(t) - \xi_{p'l'}(t + \Delta t)\right)} e^{-j2\pi f_{\max}\cos\gamma_R\Delta t} \tag{21}$$

where

$$\xi_{p'l'}(t + \Delta t) \approx \xi_{p'l'}(t) - v_R\Delta t\cos\beta_T\cos(\alpha_{\text{LoS}}(t) - \gamma_R) \tag{22}$$

—In the case of NLoS components

$$R^{\text{NLoS}}_{pl,p'l'}(\delta_T, \delta_R, t, \Delta t) = \frac{1}{\delta(i-1)K + 1} \times \int_{-\beta_m}^{\beta_m}\int_{-\pi}^{\pi} e^{-j2\pi f_{\max}\cos(\alpha_{n_i} - \gamma_R)\cos\beta_{n_i}\Delta t}$$

$$e^{-j\frac{2\pi}{\lambda}\left(\xi_{pn_i}(t) + \xi_{n_il}(t) - \xi_{p'n_i}(t + \Delta t) - \xi_{n_il'}(t + \Delta t)\right)} f(\alpha_{n_i}(t))\, f(\beta_{n_i}(t))\, d\alpha_{n_i}(t)\, d\beta_{n_i}(t) \tag{23}$$

where

$$\xi_{n_i l'} \left(t + \Delta t \right) \approx \xi_{n_i l'} \left(t \right) - v_R \Delta t \cos \left(\alpha_{n_i} \left(t \right) - \gamma_R \right) \cos \beta_{n_i} \left(t \right) \qquad (24)$$

5   Simulation Model

The theoretical model assumes an infinite number of scatterers, which prevents software and/or hardware implementation. Here, it needs to design simulation models with a finite number of scatterers which capturing non-stationary channel characteristics as accurate as possible. The corresponding simulation model of HAP-MIMO channel model is given by

$$\tilde{h}_{i,pl} \left(t \right) = \delta \left(i - 1 \right) \tilde{h}_{i,pl}^{\text{LoS}} \left(t \right) + \tilde{h}_{i,pl}^{\text{NLoS}} \left(t \right) \qquad (25)$$

where

$$\tilde{h}_{1,pl}^{\text{LoS}} \left(t \right) = \sqrt{\frac{K}{K+1}} \times \exp \left\{ -j2\pi \xi_{pl} \left(t \right) / \lambda + j2\pi f_{\max} \cos \left(\alpha_{\text{LoS}} \left(t \right) - \gamma_R \right) t \right\} \quad (26)$$

$$\tilde{h}_{i,pl}^{\text{NLoS}} \left(t \right) = \sqrt{\frac{\Omega_i}{\delta \left(i - 1 \right) K + 1}} \frac{1}{\sqrt{M_E N_A}} \sum_{m=1}^{M_E} \sum_{n=1}^{N_A} g_{mn} \exp \left(j\varphi_{mn} \right)$$
$$\times \exp \left\{ -j2\pi / \lambda \left(\xi_{pmn} \left(t \right) + \xi_{mnl} \left(t \right) \right) + j2\pi f_{\max} \cos \left(\alpha_m \left(t \right) - \gamma_R \right) \cos \beta_n \left(t \right) t \right\}$$
$$(27)$$

The phases φ_{mn} are random variables uniformly distributed in the interval $[-\pi, \pi)$. Here, we consider a nonisotropic scattering environment, we will use the method of equal areas (MEA) to calculate azimuth angles and elevation angles for the simulation model. By using the MEA, the $\beta_m \left(t \right)$ can be determined by

$$\beta_m \left(t \right) = \frac{2\beta_{\max}}{\pi} \arcsin \left(\frac{2 \left(m - 1/4 \right)}{M_E} - 1 \right) \qquad (28)$$

By applying the MEA, the $\alpha_n \left(t \right)$ can be determined by using numerical root-finding techniques [17]

$$\frac{n - 1/4}{N_A} - \int_{-\pi}^{\alpha_n(t)} f \left(\tilde{\alpha}_n \left(t \right) \right) d\tilde{\alpha}_n \left(t \right) = 0 \qquad (29)$$

6   The Capacity of the Non-stationary HAP-MIMO Channel

In this section, we defined the non-stationary HAP-MIMO channel capacity and demonstrate the utility of the proposed model. The MIMO capacity can be obtained by [6]

$$C = \log_2 \det \left(I_{N_R} + \left(\frac{SNR}{N_T} \right) \mathbf{H} \left(t \right) \mathbf{H} (t)^H \right) \ bps/Hz \qquad (30)$$

where $\mathbf{H}(t)$ is $N_R \times N_T$ matrix of complex faded channel gains, I_{N_R} is the identity matrix of size N_R, SNR corresponds to the average signal-to-noise ratio at the input of the receiver, $(\cdot)^H$ denotes the complex conjugate transpose operator, and $\det(\cdot)$ denotes the matrix determinant.

The Ricean channel matrix $\mathbf{H}(t)$ can be expressed as

$$\mathbf{H}(t) = \sqrt{\frac{K}{K+1}}\mathbf{H}_{\mathrm{LoS}}(t) + \sqrt{\frac{1}{K+1}}\mathbf{H}_{\mathrm{NLoS}}(t) \tag{31}$$

where $\mathbf{H}_{\mathrm{LoS}}(t)$ and $\mathbf{H}_{\mathrm{NLoS}}(t)$ denotes the channel matrix of LoS components and NLoS components with the size $N_R \times N_T$, respectively, K is Ricean factor. The LoS components $\mathbf{H}_{\mathrm{LoS}}(t)$ can be given by

$$\mathbf{H}_{\mathrm{LoS}}(t) = \begin{bmatrix} h_{1,11}^{\mathrm{LoS}}(t) & h_{1,12}^{\mathrm{LoS}}(t) \\ h_{1,21}^{\mathrm{LoS}}(t) & h_{1,22}^{\mathrm{LoS}}(t) \end{bmatrix} \tag{32}$$

The NLoS components $\mathbf{H}_{\mathrm{NLoS}}(t)$ can be given by [18]

$$\mathrm{vec}\,(\mathbf{H}_{\mathrm{NLoS}}(t)) = \mathbf{R}_{\mathrm{NLoS}}^{1/2}(t)\,\mathrm{vec}\,(\mathbf{H}_w) \tag{33}$$

where $\mathrm{vec}\,(\cdot)$ denotes matrix vectorization, $\mathbf{R}_{\mathrm{NLoS}}(t)$ is the $N_R N_T \times N_R N_T$ correlation matrix associated with the NLoS components, $\mathbf{R}_{\mathrm{NLoS}}^{1/2}(t)$ is the square root of $\mathbf{R}_{\mathrm{NLoS}}(t)$ that satisfies $\mathbf{R}_{\mathrm{NLoS}}^{1/2}(t)\mathbf{R}_{\mathrm{NLoS}}^{H/2}(t) = \mathbf{R}_{\mathrm{NLoS}}(t)$, and $\mathbf{H}_w$ is a $N_R \times N_T$ stochastic matrix with independent identically distributed zero mean complex Gaussian entries. Considering a 2×2 HAP-MIMO channel, $\mathbf{R}_{\mathrm{NLoS}}(t)$ can be given by

$$\mathbf{R}_{\mathrm{NLoS}}(t) = \begin{bmatrix} R_{11,11}^{\mathrm{NLoS}}(t) & R_{11,21}^{\mathrm{NLoS}}(t) & R_{11,12}^{\mathrm{NLoS}}(t) & R_{11,22}^{\mathrm{NLoS}}(t) \\ R_{21,11}^{\mathrm{NLoS}}(t) & R_{21,21}^{\mathrm{NLoS}}(t) & R_{21,12}^{\mathrm{NLoS}}(t) & R_{21,22}^{\mathrm{NLoS}}(t) \\ R_{12,11}^{\mathrm{NLoS}}(t) & R_{12,21}^{\mathrm{NLoS}}(t) & R_{12,12}^{\mathrm{NLoS}}(t) & R_{12,22}^{\mathrm{NLoS}}(t) \\ R_{22,11}^{\mathrm{NLoS}}(t) & R_{22,21}^{\mathrm{NLoS}}(t) & R_{22,12}^{\mathrm{NLoS}}(t) & R_{22,22}^{\mathrm{NLoS}}(t) \end{bmatrix} \tag{34}$$

7   Numerical Results

The parameters for the following numerical results and analysis are listed here: $H_T = 20\,\mathrm{km}$, $f = 2$ GHz, $K = 6$, $\theta_T = \pi/3$, $\beta_T = \pi/6$, $\kappa = 10$, $\beta_{\max} = \pi/6$, $\theta_R = \pi/4$, $\varphi_R = \pi/3$, $v_R = 8\,\mathrm{m/s}$, $\gamma_R = \pi/4$, $N_A = 30$, $M_E = 15$, $R_1(t_0) = 75$ m, $R_2(t_0) = 80$ m, $\mu_1(t_0) = \mu_2(t_0) = \pi/3$.

Figure 4 shows the time-variant correlation function for tap 1 and tap 2 at different time t. From the Fig. 4, we can easily see that the tap 1 has a higher correlation compared with the tap 2 because of the dominant LoS components.

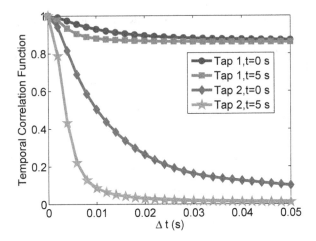

Fig. 4. Time-variant correlation function of different taps.

Figure 5 shows a comparison which is made against temporal correlation function of Rayleigh channel for theoretical model and simulation model. From the Fig. 5, we can see that the simulation model align well with theoretical model, demonstrating the correctness of theoretical derivations and simulations.

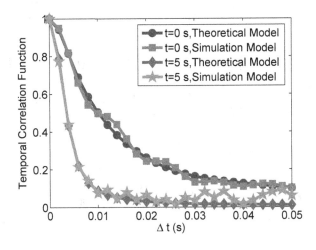

Fig. 5. A comparison which is made temporal correlation function between the theoretical model and simulation model.

Figure 6 shows the spatial correlation function for tap 1 and tap 2 at different time t. From the Fig. 6, we can easily see that the tap 1 has a higher correlation compared with the tap 2 because of the dominant LoS components.

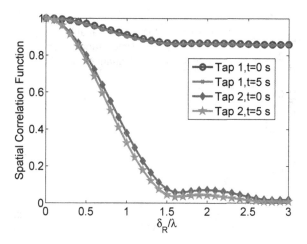

Fig. 6. Spatial correlation function of different taps.

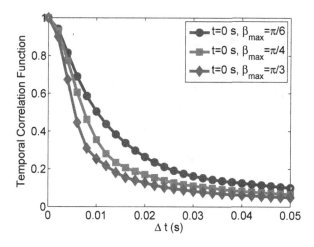

Fig. 7. Comparison between different maximum elevation angles $\beta_{\max}$.

Figure 7 describes the time-variant correlation function of tap 2 for different $\beta_{\max}$. From the Fig. 7, we can easily see that the correlation function is effected by the elevation angle of scatterers.

Figure 8 describes the correlation function of tap 2 of time-variant channel model and non-time-variant channel model. From the Fig. 8, we can see that the correlation function of non-time-variant channel is not changing at different t. So the non-time-variant couldn't describe the non-stationary property of HAP-MIMO channel.

Figure 9 describes the capacity of tap 1 and tap 2 with different antenna elements at transmitter and receiver at time $t = 5$ s. From the Fig. 9, we can see

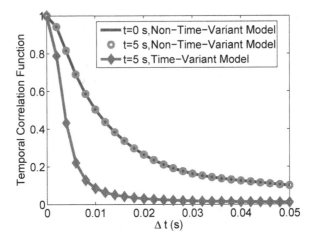

Fig. 8. Comparison between time-variant and non-time-variant channel.

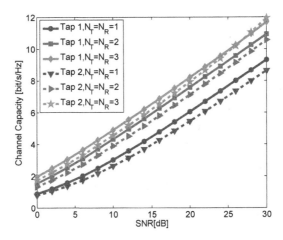

Fig. 9. Channel capacity for different taps.

that the channel capacity increases as the number of antenna elements increases. So the HAP-MIMO channel could improve the channel capacity of HAP-MIMO communication systems.

Figure 10 describes the capacity of tap 2 with different antenna elements at transmitter and receiver at time $t = 0$ s and $t = 5$ s. From the Fig. 10, we can see that the channel capacity different at different time t. The traditional channel model couldn't describe the time-variant property of non-stationary channel. And it is important to investigate the non-stationary HAP-MIMO channel.

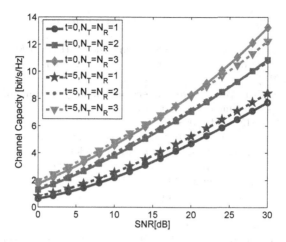

Fig. 10. Channel capacity for time-variant channel.

8 Conclusion

A theoretical non-stationary 3-D multi-cylinder model has been proposed in this paper. The space-time correlation function of the proposed 3-D HAP-MIMO channel has been proposed. In addition, we have proposed a corresponding simulation model. We have investigated the capacity of spatially and temporally correlated HAP-MIMO channel achieved with ULAs.

Acknowledgments. This research was supported by the National Natural Science Foundation of China under Grands # 91438113.

References

1. Husni, E.M., Razali, R., Said, A.M.: Broadband communications based on high altitude platform systems (HAPS) for tropical countries. In: International Symposium on Signal Processing and its Applications (ISSPA), Kuala Lumpur, Malaysia, vol. 2, pp. 517–520, August 2001
2. Ohmori, S., Yamao, Y., Nakajima, N.: The future generations of mobile communications based on broadband access technologies. IEEE Commun. Mag. **38**(12), 134–142 (2000)
3. He, C., Zhu, H.-W., Wu, G., Miura, R., Hase, Y.: Dynamic resource assignment for stratospheric platform communication system with multi-beam antenna. In: IEEE International Conference on Communications (ICC 2001), Helsinki, Finland, June 2001
4. Grace, D., Mohorcic, M., Capstick, M.H., Pallavicini, M.B., Fitch, M.: Integrating users into the wider broadband network via high altitude platforms. IEEE Wirel. Commun. **12**(5), 98–105 (2005)
5. Karapantazis, S., Pavlidou, N.: Broadband communications via high-altitude platforms: a survey. IEEE Commun. Surv. Tutorials **7**(1), 2–31 (2005). First Qtr

6. Telatar, I.E.: Capacity of multi-antenna Gaussian channels. Eur. Trans. Telecommun. **10**(6), 585–595 (1999)

7. Michailidis, E.T., Kanatas, A.G.: Three-dimensional HAP-MIMO channels: modeling and analysis of space-time correlation. IEEE Trans. Veh. Technol. **59**(5), 2232–2242 (2010)

8. Zajic, A.G., Stuber, G.L.: Three-dimensional modeling, simulation, and capacity analysis of space-time correlated mobile-to-mobile channels. IEEE Trans. Veh. Technol. **57**(4), 2042–2054 (2008)

9. Zajic, A.G., Stuber, G.L., Pratt, T.G., Nguyen, S.T.: Wideband MIMO mobile-to-mobile channels: geometry-based statistical modeling with experimental verification. IEEE Trans. Veh. Technol. **58**(2), 517–534 (2009)

10. Dovis, F., Fantini, R., Mondin, M., Savi, P.: Small-scale fading for high-altitude platform (HAP) propagation channels. IEEE J. Sel. Areas Commun. **20**(3), 641–647 (2002)

11. King, P.R., Evans, B.G., Stavrou, S.: Physical-statistical model for the land mobile-satellite channel applied to satellite/HAP MIMO. In: Proceedings of the 11th European Wireless Conference, Nicosia, Cyprus, vol. 1, pp. 198–204, April 2005

12. Mahmoud, S.S., Hussain, Z.M., O'Shea, P.: A geometrical-based microcell mobile radio channel model. Wirel. Netw. **12**(5), 653–664 (2006)

13. Yuan, Y., Wang, C.X., He, Y., Alwakeel, M.M., Aggoune, E.H.M.: 3D wideband non-stationary geometry-based stochastic models for non-isotropic MIMO vehicle-to-vehicle channels. IEEE Trans. Wirel. Commun. **14**(12), 6883–6895 (2015)

14. Schubert, F.M., Jakobsen, M.L., Fleury, B.H.: Non-stationary propagation model for scattering volumes with an application to the rural LMS channel. IEEE Trans. Antennas Propag. **61**(5), 2817–2828 (2013)

15. Abdi, A., Kaveh, M.: A space-time correlation model for multielement antenna systems in model fading channels. IEEE J. Sel. Areas Commun. **20**(3), 550–560 (2002)

16. Parsons, J.D., Turkmani, A.M.D.: Characterization of mobile radio signals: model description. IEE Proc.-I **138**(6), 549–556 (1991)

17. Mardia, K.V., Jupp, P.E.: Directional Statistics. Wiley, New York (1999)

18. Shiu, D.-S., Foschini, G.J., Gans, M.J., Kahn, J.M.: Fading correlation and its effect on the capacity of multielement antenna systems. IEEE Trans. Commun. **48**(3), 502–513 (2000)

Sparse Representation and Fusion Process in Space Information

Double Layer LEO Satellite Based "BigMAC" Space Information Network Architecture

Keke Zhang[1,2,3(✉)], Lei Xia[2], Shengyu Zhang[2,3], Chaoming Si[2], and Shilong Zhou[2]

[1] Shanghai Institute of Microsystem and Information Technology, CAS, Shanghai, China
13816686945@139.com
[2] Institute of Winano Satellites, Innovation Academy for Microsatellites of CAS, Shanghai, China
[3] University of Chinese Academy of Sciences, Beijing, China

Abstract. Space information network (SIN) is not only the key element to contact everything all over the earth, but also the important part of the information highway in the future. Architecture design is the basement of the whole space information network, in order to insure the access of the existing various types of satellites, aircrafts and ground facilities, while taking into account the cost of satellite and system performance, we design a double layer LEO satellite based "BigMAC" space information network architecture in this paper, which is compatible to the various existing satellite systems. The simulation results shows that the architecture is feasible and can provide others some construction of ideas in the future.

Keywords: Space Information Network · Architecture · Low orbit small satellite

1 Introduction

With the constant progress of science, rapid development of information technology and its extensive applications in national defense and civilian use, new reform of space scene characterized by informationization is in the ascendant around the world. It is urgent to realize space-aeronautics incorporation in national defense and civilian use and various countries reach a consensus to capture advantages of informatization in the space field [1]. It screams for studying the system and critical technology for space information network and establishing network system of space-aeronautics incorporation by virtue of ground internet, internet of things and satellite constellation [2].

The study on space information network architecture is the footstone of constructing space information network. The global information fence architecture model, Command Control Constellation Net (C2Cnet) [3] and interplanetary internet architecture model [4, 5] have already been proposed in European and American countries. Recently, commercial satellite network develops rapidly [6], including space information network architecture. For example, the American Oneweb company spent $1 billion USD creating low orbit communication satellite constellation. China proposed the concept of "space information network" during the period of "11th Five-year Plan" and has acquired some research achievements by using some critical technology in space information network construction for a dozen years of studies. Studies may be concentrated on

© Springer Nature Singapore Pte Ltd. 2017
Q. Yu (Ed.): SINC 2016, CCIS 688, pp. 217–231, 2017.
DOI: 10.1007/978-981-10-4403-8_19

routing and network management technology [7–10], but there are a few studies on space information network architecture beginning with in-depth comprehension on overall and strategic demand planning in China.

Relative to the traditional large-scale satellite communication system, the space information network architecture design based on low-cost small satellites is applied in the paper. On the one hand, it safeguards seamless access of various aircrafts, spacecraft and ground terminals in space, sky and ground. On the other hand, the design mode of small satellites is applied to reduce construction costs of space information network and it has the important significance on speeding up transforming technological demonstration of space information network to engineering implementation, satisfying development tendency of space-aeronautics incorporation in future information network, and further developing the role of space assets in national politics, economy, military and civilian use construction.

2   Design of Space Information Network Architecture

2.1   Analysis of On-orbit Satellites

The exiting spacecraft apply the typical stove-piped management, so it will be confronted with challenges to access on orbit system. Firstly, it is necessary to analyze on orbit system. Until 2016, the distribution of on orbit satellites has had no great changes, distributing in LEO and GEO. Earth observation satellites mainly concentrate on LEO. Communication and navigation satellites are mainly distributed in GEO and MEO.

It can be observed from Fig. 1 that for LEO satellites, majorities of them are distributed in the special scale ranging from 400 km to 1400 km and mainly concentrate on 400 km and 800 km. In the range, there are nearly 500 satellites, exceeding 40% of the total on orbit satellites. Earth observation satellites exceed 80% of all earth observation satellites.

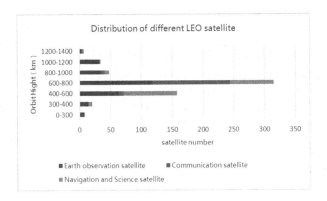

Fig. 1. The main purpose of different low orbit satellites

2.2   "BigMAC" Architecture Design

Space information network serves for space information nodes, so space information network design considers access of space nodes and stability of data links as input conditions.

Overseas broadband multimedia satellites serving for earth communication nodes may apply the following space structure, including:

- single layer LEO architecture: Iridum (66 satellites), Teledessic (288 satellites), SkyBridge (80 satellites);
- single layer MEO architecture: Odyssey (12 satellites), O3b (12 satellites);
- single layer GEO architecture: Antrolink (9 satellites), Cyberstar (3 satellites), iSky (2 satellites);
- double layer GEO + MEO architecture: Spaceway (16 GEO satellites + 20 MEO satellites);
- double layer GEO + LEO architecture: Celestri (9 GEO satellites + 63 LEO satellites).

At present, China constructs a batch of satellite system according to application fields, such as Compass, high resolution remote sensing system, communication satellite net, relay satellite system, weather satellite system and environmental satellite system, etc., which are improving constantly. However, it is relatively independent between systems. As a result, in order to construct the integrated information network, the best solution means to both consider the existing system and future development.

Therefore, the construction of space information network architecture firstly should apply the existing on orbit satellites to space network to be constructed and provide customizable and extensible architectural features for future satellites.

The difference between space information network and satellite communication network means that service objects are space information nodes. Based on the property of space information network, the "BigMAC" space information network architecture is proposed (Fig. 2).

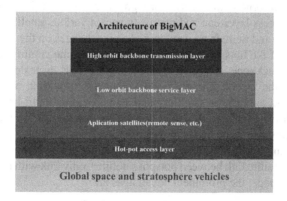

Fig. 2. "BigMAC" space information network architecture

"BigMAC" network architecture: forms the main architecture providing space information services for application satellites and other space nodes by constructing a real-time, stable and interconnected backbone core layer and offering a hotspot access layer covering the whole airspace. Moreover, it constitutes the space information system with application nodes for perception, calculation, transmission, storage and control management.

1. Backbone core layer

It is composed of space nodes with space routing, on orbit computing and information management. Composition of satellites in the layer can be divided into different configurations.

The backbone core layer includes LEO backbone core nodes of 1400 km orbit and backbone nodes in middle and high orbits. The higher orbits in LEO are applied to deploy core constellation, to realize full coverage of nodes on the hotspot layer on the basis of the limited satellites, and to realize low-delayed space information network with low comprehensive costs.

Firstly, backbone core layer is responsible for backbone transmission, has the ability of high-capacity and high-rate data transmission, and is equipped with space routing function and network management function, etc. Moreover, it can utilize the higher performance and provide on orbit computing and high-capacity storage, etc.

Meanwhile, with the business development, for some space information business with high-capacity data, high-rate transmission, and low-delayed demand, the middle and high orbit backbone core nodes can be established gradually. Meanwhile, backbone core layer to be constructed can be compatible with the existing communication system, improving system availability and performance.

Main features of "BigMAC" space information network architecture mean to remain expandability for compatibility of the existing system and future network development.

2. 350 km hotspot access layer

The satellites in this layer mainly provide service access functions of space equipment (including airborne equipment). Through the analysis of the existing satellites, 350 km hotspot access layer can and passively accept various satellite data more than 400 km effectively and there is no need to make any changes on the existing satellites. The access node of the hotspot layer maintains the real-time communication with the core layer of the backbone. In principle, the hotspot access layer does not have inter-satellite links in the layers. The hotspot layer nodes, providing access to different protocol systems of application satellites, and also providing multi-hop and multi-user connection, with the exchange function, convey the information in the space service node to the backbone satellite nodes in real-time.

Take a large number of hotspot access to the satellite deployment in the lower orbit, there is no inter-satellite link between hotspot satellites, only link with the business node and the backbone node, the prominent advantages are:

- Low-cost micro-satellites can be made as the node of the hotspot access layer, which can not only reduce the construction cost of the system, but also can shorten the cycle of technology update;

- Facilitate the access of the service nodes that are widely distributed in the higher orbital space, while reducing its transmission distance with the adjacent space and stratospheric aircraft, and facilitating the access of service nodes (Fig. 3).

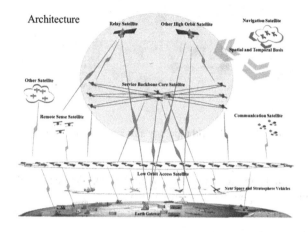

Fig. 3. "BigMAC" architecture diagrammatic sketch and topology

Main features of "BigMAC" space information network architecture mean to remain expandability for compatibility of the existing system and future network development. The backbone layer network satellite carries the data transmission task of the spatial information network. It is supposed to carry 10 Gbps laser communication load and 3 Gbps microwave communication load. The two communication rates are used to carry different service types such as short message and other low data business. After being collected by the hotspot layer satellites, it can adopt the microwave low-rate transmission mode to carry on the service. For the business type with weak timeliness, it can make information aggregation through the hotspot layer, and then used to forward by the backbone layer satellite. For the high-speed, large-data traffic flow mode such as video streaming, the backbone layer can be transmitted by using laser high-speed communication mode. As a result, the entire architecture can address multiple business models.

2.3 Topology Structure of "BigMAC"

The topology of "BigMAC" spatial information network is a kind of hybrid topology with semi-distributed center structure (Fig. 4).

The network topology has the following topology characteristics:

- In the "BigMAC" network, the static topology at any one time can guarantee the connectivity of any two nodes;
- At the same time, in the network topology, most of the nodes maintain an adequate redundant path in most of the time. The redundancy degree of the link increases with the number of nodes in the network, greatly improving the reliability and high fault-tolerant capability;

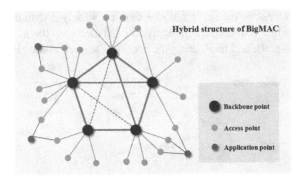

Fig. 4. The hybrid topology structure of "BigMAC"

- The network organized by the backbone nodes is viewed as the network structure, any two nodes can be interconnected through the link. While the link between nodes is more fixed or equipped with time regularity;
- The nodal degree of the backbone node and the access node is changing dynamically, existing a high degree of node degree change;
- There is no link between the access nodes, and the access node is only connected with the backbone node and the application node, and the access node can also establish links with multiple backbone nodes and multiple application nodes at the same time;
- The application node is only linked with the access node, and it can also link with a plurality of access nodes at the same time;
- SuperNodes exist in the "BigMAC" network, the backbone node, as a super node with a strong performance and core role, can store a large number of information of other nodes, and achieve complex operations, therefore, only the backbone core node can achieve the realization of the space routing algorithm under the high dynamic situation in the real time, the function of on-orbit management of space network. So "BigMAC" network structure is also showing the characteristics of the tree structure in the profile. The core layer of the backbone reflects a more important role. For the large data computing services, once the backbone nodes can not be carried by the satellite, we can consider routing to the nearest ground station for auxiliary operations, or making auxiliary operation by the peripheral spare backbone node satellites with the help of spatial information network.

Main advantages of "BigMAC" semi-distributed hybrid topological structure are shown as follows:

Firstly, it satisfies the demands of connecting with all on orbit systems by space information network and realizing full-day and full-airspace intelligibility and interconnection of space nodes, namely the topological structure is equipped with the good connectivity;

Secondly, space topological structure is equipped with highly dynamic behaviors, but topological structure remains intelligibility and interconnection of two arbitrary nodes and some path redundancy at any static time in highly dynamic conditions, namely the topological structure is equipped with good stability;

Thirdly, the topological structure makes a distinction between the important essential and the lesser one, simplifies systematic complexity, and improves important degree of backbone nodes, thus it can conduct system extension based on backbone nodes. Moreover, with the node increase and system expansion, it greatly improves reliability and optimal space, while increasing systematic complexity infinitely, namely the topological structure is equipped with good expansibility;

In addition, with the core of backbone nodes, the topological structure can solve some risk and safety problems by setting up stratified backbone nodes, setting up different access rights and data encryption, and remaining some redundancy to control. When backbone nodes are destroyed, it only will increase system risks or reduce some performance, but won't result in overall functional fault and failure, namely the topological structure is equipped with good safety.

The topological structure applied by "BigMAC" is the optimal design result by combining with global superiority, high dynamics, business differences and different effect requirements of space information network and is suitable for gradual and staging on orbit construction of space information network. However, the construction design also has some design defects. In order to safeguard non-modified adaption to the existing satellite system, a layer of low orbit access satellites are designed with more quantity and lower orbit height. Satellite orbit on the layer has relatively short lifetime, so it is necessary to maintain regularly. Meanwhile, satellites on the backbone layer still have a large difference on the calculated quantity and storage space by comparing with big traditional satellites, due to apply low orbit small satellites. "BigMAC" architecture is the initial construction of space information network. It is thought that in the future satellite development, subsequent successful satellites will be gradually equipped with the functions of access satellites and application satellites with the constant improvement of star ability and constant maturity of star routing technology and laser communication technology. The entire network architecture also will present the semi-incompact self-organizing network mode gradually and control the entire network through SuperNodes on the higher orbit.

2.4 Information Flow of "BigMAC" Space Information Network

Architecture information flow of "BigMAC" system is as follows (Fig. 5):

- Users' task distribution phase, the user will make the required command, for example, monitoring a certain area, tracking certain types of objects, access to the satellite through the terminal by using LEO hotspot in the information collection layer of 350 km orbit; and then passed them to backbone node satellites layer in the 1400 km orbit height, or can also be directly delivered to the backbone pot of satellite nodes with the help of high-performance equipment;

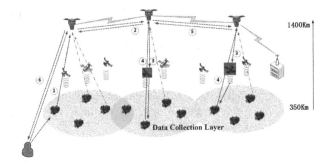

Fig. 5. Information flow of "BigMAC"

- After the network satellite in the backbone network receives the user's task request, it intelligently judges and decomposes the task, finds the target function star of the task demand, and forwards the message to the target function star using the routing function of spatial information network;
- Target function satellites, such as earth observation satellites, remote sensing satellites, and electronic reconnaissance satellites, can operate the corresponding task operation according to the received tasks, collect relevant data, access to the network satellites in the access layer located below it, and send the collected data to the network satellite in the access later;
- After receiving the data, the network satellite in the access layer will send the data to the accessible network satellite in the backbone layer at the current time;
- After the network satellite in the backbone layer gets the relevant information, it analyzes the data packet, and finds out the internet location information of the destination user, and returns the data packet. Since the spatial information network belongs to the high-speed dynamic topology, the spatial information network configuration in process 5 and process 2 changes as shown in Fig. 6, which need to re-find the optimal routing path, and return the information through the optimal routing path;

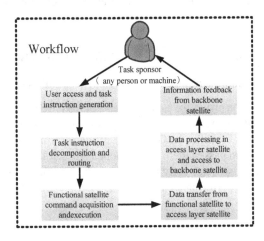

Fig. 6. The route chart of "BigMAC"

- When the data arrives at the network satellites in the backbone layer that can cover the network location, the network satellite in the backbone layer will transmit the information to the user.

From the work process, it can be seen that the most core two key technologies in the initial construction program of two-tier structure of spatial information network are: rapid dynamic access technology among various types of nodes, followed by is high-efficient routing technology in the backbone layer of spatial information network. The two technologies form the foundation and guarantee of spatial information network.

3   Performance Analysis of Space Information Network

3.1   Visibility Analysis Between Different Layer

The system has designed the interconnection of the links among the 15 backbone nodes of 1400 km orbit in real time, and the lower layer satellites can be completely covered at any time, thus the 144 access nodes in the 350 km orbits need to be connected to the existing system and other equipment at the same time (Fig. 7).

1. Visibility analysis of 1400 km orbit backbone layer satellites

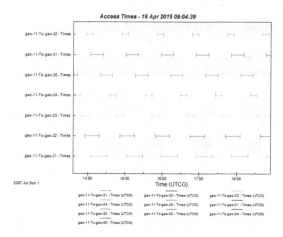

Fig. 7. Analysis of 1400 km orbit satellites

From the visibility analysis of the 1400 km layer, we see that there is no inter-satellite link among the same orbital plane of the 15 constellations, which only exists in the different orbital planes, and any backbone star can be guaranteed to connect with at least one backbone star at any time.

2. Visibility analysis between 1400 km backbone layer and 350 km hotspot layer

From the visibility analysis between the 1400 km layers and 350 km layers, it can be seen that the hot star can be guaranteed to be connected with at least one backbone star at the real-time, and the connection between the hot star and the backbone star is switched over time, including the switch of backbone satellite in inter-satellite and the switch in the orbit Plane of backbone satellite (Fig. 8).

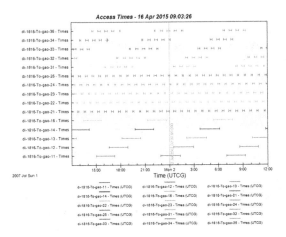

Fig. 8. Analysis between 1400 km and 350 km satellites

3. The coverage analysis between hotspot layer and other satellites

The access efficiency of the access layer of the orbit satellites is mainly related to the orbital height of the target satellite and the ground beam angle. The beam angle of the general satellite-to-ground propagation is 75° and the 60° beam angle is guaranteed to be available. The following figure is the access network's coverage capability on the altitude of each orbit and the target satellites with beam angle capability (Table 1).

Table 1. Coverage analysis with different orbit height and beam angle

User satellite orbit height	500	600	700	800	900	1000
75° beam angle	61.86%	100%	100%	100%	100%	100%
65° beam angle	15.94%	40.65%	71.81%	100%	100%	100%
55° beam angle	6.10%	16.72%	31.23%	48.82%	68.85%	90.82%

According to the Fig. 9, not all the in-orbit satellites can realize real-time access, and the closer the target orbit altitude is close to the satellite orbit in the access-level, the smaller the possibility of real-time access is. In addition, some on-orbit satellites also have a narrow band antenna, so, if this part of the satellites want to access, they can only be accessed by attitude maneuver. Taking into account that most of the on-orbit satellites have maneuver capability, the Fig. 10 shows the desired beam angle that the target satellite wants to access to the information acquisition network to determine the required maneuver capability based on its own beam angle.

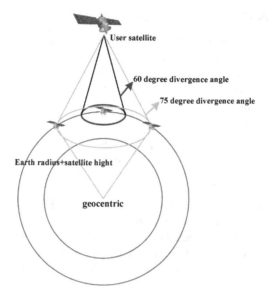

Fig. 9. User satellites access chart

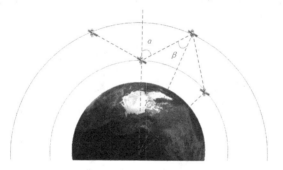

Fig. 10. Access requirements of beam angle

In the figure, $\alpha = \beta + \gamma$, α is the beam angle required for the access satellite to receive, β is the beam angle of the launching of the target satellite, γ is the corresponding earth angle of the two satellites, and also half of the corresponding earth satellite between the neighboring two satellites of the access constellation.

Each orbit plane of the access layer constellation has 16 satellites, so $\gamma = 360°/16/2 = 11.25°$. According to the different target orbit heights, we can calculate the β angle, that is, the beam angle, which is required for the target of each orbital altitude to cover the access layer in real time (Table 2).

Table 2. The beam angle requirements for real-time access

User satellites orbit height	Beam angle required
400	82.75°
500	78.51°
600	74.39°
700	70.44°
800	66.66°
900	63.09°
1000	59.74°

Considering the ground beam angle of the existing satellites, the required attitude maneuver angle can be determined. If the existing satellite in the 500 km high-altitude uses a narrow beam antenna with a beam angle of 20°, when it can realize the real-time access, the required maneuver angle is $78.51 - 20 = 58.51°$. If the existing satellite in the 500 km high-altitude uses wide-beam antenna, with the beam angle of 60°, when it realize the real-time access, the required motor angle is $78.51 - 60 = 18.51°$.

3.2 Optimal Path Analysis of System Link

The LEO backbone layer is designed with minimalist design method, the minimalist design schematic is shown in Fig. 11.

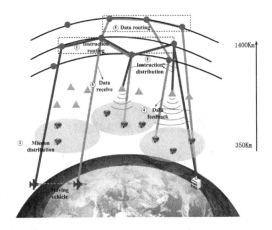

Fig. 11. Minimalist design for LEO backbone layer

Notes: The figure between two stars stands for the communication link and its fitted value, including distance, delay and jamming intensity, etc (Fig. 12).

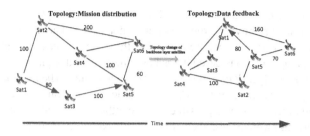

Fig. 12. The topology change of LEO backbone layer

Dijkstra algorithm is applied to solve the worst communication link to solve and analyze. By calculating and analyzing the network topology and optimal routing path in every second of 86400 s on April 21st, 2015, maximal optimal path between two backbone network satellites at anytime for space information network of 15 backbone network satellites can be obtained as 49669.6 km. It is the optimal transmission path at 60th second from satellite node 2 to satellite node 9. Information can arrive in the target node from origin node at most for 7 jumps (Table 3).

$$t_{\text{ISL-Delay}} = (\sum D_{\text{ISL}})/C = (49669.6 + 1400 \times 2)/(3 \times 10^5) = 0.174\text{s}$$

Table 3. Distance analysis table between any two stars

21 Apr 2015 10:29:13			
Chain-Chain1			
gao-11 to gao-21		gao-21 to gao-32	
Time (EpSec)	Range (km)	Time (EpSec)	Range (km)
448	8911.186598	0	6765.971778
508	8362.926400	60	7201.091533
...	...	...	...
21912	2741.085125	29772	8405.327559
...	...	...	...
86400	834.722909	85455	8886.760824

Under the worst situation, inter-star transmission delay is 0.175 s. Relative to traditional relay satellite system, the optimal transmission path distance is 72000 km. the minimal transmission delay is 0.24 s, thus the backbone network constituted by 15 satellites can distribute orders and pass back information more rapidly and effectively. Because the "BigMAC" system structure mainly adopts the low-orbit satellite to design the backbone layer and the access layer, the change rate of the dynamic topology in the low-orbit satellite constellation is fast, which may result in the interruption of the communication link in the network. Broken link can be divided into the following categories: when the backbone layer nodes transmit information, due to the routing interrupt caused by topology changes, the laser communication load of the satellite in the backbone layer makes the interruption of the communication link caused by high-speed mobile or interference, and the access satellite forms the interruption of the communication link because of removing from the satellite node control area in the current backbone layer since the topology changes. For the above types of link disruption, "BigMAC" has made full considered in the beginning of the design: First, routing interrupt in the backbone layer due to the topology changes, since when the backbone layer satellite is designed, each satellite should keep at least one redundant communication link at the same time, hence, it is possible to select other redundant communication link for information routing once the current transmission path is interrupted; secondly, when the problem of communication interruption is appeared in the high-speed mobile state of the laser communication, in the research of the laser communication key technology of the spatial information network, we have carried out detailed design and consideration, under normal circumstances, it can meet the communication capture under the movement of satellite in the backbone layer; if the occurrence of communication link disruption is happening, you can also re-capture networking in a very short period of time; Finally, the access layer satellite caused the upper layer and lower layer communication disruption due to it is removed out of the current satellite node control area in the backbone layer for the topology change, and this has been carried out detailed demonstration in the dynamic access of the spatial information network and key technology research in the mobile switching; the network has the corresponding access and switching strategy to protect the communication maintaining when the access layer satellite changes their location (Fig. 13).

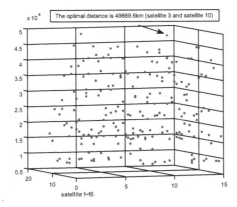

Fig. 13. The optimal path of maximum distance between any two stars

4   Conclusion

"BigMAC" space information network architecture scheme, on the one hand, provides the access platform for various on orbit spacecraft, so that space nodes won't do things in their own way. Multi-source information can be integrated and disposed rapidly. On the other hand, in the future development, "BigMAC" space information network architecture is equipped with good expandability. With network performance analysis, space network nodes can be better integrated into the entire network system, while improving respective performance. Network performance will be improved gradually. Network services also will be optimized, finally forming integrated network of air, sea and land. It will provide more real-time, effective, convenient and high-quality services for both military applications and civilian use demands.

References

1. Chang, Q., Li, X., He, S.: Dicussion about China space information network development. J. Telemetry Track. Comm. **36**(1), 1–10 (2015)
2. Min, S.: Conception of China space information network. Spacecr. Eng. **22**(5), 1–14 (2013)
3. Sweet, N., Kanefsky, S.: The C2 Constellation a US air force network centric warfare program. C2 Constellation a US Air Force Network Centric Warfare Program (2004)
4. Farserotu, J., Prasad, R.: A survey of future broadband multimedia satellite systems, issues and trends. IEEE Commun. Mag. **38**(6), 128–133 (2000)
5. Surhone, L.M., Tennoe, M.T., Henssonow, S.F.: InterPlaNet. Betascript Publishing (2010)
6. Martínez, P., Francisco, O.U.G., Gustavo, A.M.P., et al.: OneWeb: plataforma de adaptación de contenidos web basada en las recomendaciones del W3C Mobile Web Initiative. Ingenieria E Investigación **31**(1), 117–126 (2011)
7. Zhang, W., Zhang, G., Gou, L., et al.: A hierarchical autonomous system based topology control algorithm in space information network. KSII Trans. Internet Inf. Syst. **9**(9), 3572–3593 (2015)
8. Hao, X., Ma, J., Ren, F., Liu, X., Yan-tao, Z.: A kind of authentication routing protocol based on double satellite network in space information network. Comput. Sci. **38**(2), 79–81 (2011)
9. Chen, Y., Meng, X., Zhang, L.: Analysis of space information network architecture. Comput. Technol. Dev. **22**(6), 1–5 (2012)
10. Zhang, D., Liu, S.: Research on mesh-based architecture for space information network. Comput. Technol. Dev. **19**(8), 69–73 (2009)

The Assumption of the TT&C and Management for SIN Based on TDRS SMA System

Liang Zhu[✉], Huiming Huang, Jian Gao, and Bin Luo

Beijing Space Information Relay and Transmission Technology Center, Beijing, China
zhuliang3258@163.com

Abstract. In recent year, the transmission of spatial information present the characteristic of amount users, broad coverage area, multiple service, and convenient access, which make research focus for SIN (spatial information network) users in the area of intelligent TT&C (telemetry, tracking and command) and management. Based on the deep analysis of TDRSS (Tracking and Data Relay Satellite System) application modes, an assumption of TT&C and management for SIN based on the TDRS SMA (S-band Multiple Access) system is proposed. Analysis is mainly focused on the panoramic beam, system composition and user access, also the protocol architecture, communication workflow, and TDRS terminal are original designed. Compared to traditional operational mode, the proposed assumption could realize ubiquitous perception, random access, on-line management, and satellite-ground cooperation for users' application. Relevant result could be referenced in the study of TT&C and management for SIN.

Keywords: TDRSS · SMA · Measurement and control · Random access · On-line management

1 Introduction

In recent years, space information network transmission presents characteristics of large amount of users, wide coverage area, diverse business and convenient access, making the operation and management mode of space information transmission system user change from the regional time-sharing to the global pervasive, from plan-driven to collaborative driven, from offline to instant online. Therefore, it is urgent to solve the problem of intelligent telemetry and control, and network management of spatial information network users.

The relay satellite system has the characteristics of high coverage, large bandwidth and multi carrier service. The application fields cover the service objects and services of space information transmission [1]. Using relay satellite system as the backbone to build space information transmission network meets the basic characteristics of space information network, and it is also an ideal choice in line with China's national conditions [2].

The research shows that the third generations of TDRS (Tracking and Data Relay Satellite) deployed by the United States are equipped with phased array antennas, to provide users with SMA (S-band Multiple Access) service [3]. According to the

© Springer Nature Singapore Pte Ltd. 2017
Q. Yu (Ed.): SINC 2016, CCIS 688, pp. 232–243, 2017.
DOI: 10.1007/978-981-10-4403-8_20

evolution of the third generations of the United States TDRSS network access capability, SMA has been the focus of attention and continuous improvement. The backward link of the third generation of TDRSS is synthesized by terrestrial digital beam, which can form a combined panoramic beam by overlapping multiple static beams, and achieves continuous coverage of time and space. Therefore, networking by multi-satellite and designing a suitable communication system will be very suitable for building the global coverage access network of space information network, breaking through the plan-driven monitoring and control management mode, and protecting the random access requirements in condition of large amount of users. It has the ability to provide users with the service of global pervasive sense, dynamic random access and whole network online management, and it is of great significance for the intelligent telemetry and control and network management.

2   The Application Mode of TDRS SMA in the United States

So far, the United States has developed three generations of TDRSS system, serving hundreds of target users. Its target users cover space station, launch vehicles, missiles, ships, large aircraft, vehicles, unmanned aerial vehicles, civil satellites, military satellites, etc. [4]. The three generations of the United States TDRS are equipped with phased array antennas, which provide users with multiple access service of a 300Kbps forward beam and five 3Mbps backward beam. The forward channel of the SMA is mainly used to transmit tele-control commands and low-speed data to the user's spacecraft, and its backward channel is mainly used to transmit low-speed telemetry data of user's spacecraft [4]. The United States TDRS provides users with two multiple access service application mode, including week planning mode and DAS (Demand Access Service) mode. In the DAS mode of TDRS, the backward link of TDRSS-DAS uses dynamic beam tracking to track user platform, to provide users with real time and continuous data relay service, and realizes receiving and detecting random access transmission of user platform, such as the small satellite SWIFT which observes gamma-ray bursts [5].

It is shown that the United States TDRS multiple access application mode uses regional time-sharing coverage and plan-driven control mode. Because users need to send business requests through the terrestrial network, this mode is not suitable for real-time online seamless management of large amount of users, and it cannot meet the unplanned application requirements which require randomness, sudden and real time. In addition, TDRSS needs to provide a continuous transmission channel by pre-allocating fixed SMA back-to-beam resources. The inflexible resource usage and the lack of flexible and efficient protocol of interactive communication of satellite-ground command, cannot support the information exchange and network application of the space information network.

To solve the problems above, communication standard and protocol architecture of S band multiple access system for relay satellite can be designed, and space information telemetry and control management network can be constructed by using advanced mobile communication system, realizing pervasive sensing and online management to large amount of user targets, including real time receiving and processing all kinds of user location information, health status information, SOS distress information, etc. At the same time, adding a flexible user target online application based on the traditional application mode (user center initiated) is in line with the trend that any node in the space network can initiate the application in the future.

3   The Assumption of the Management System of Measurement and Control for SIN

In the multi-beam communication satellite network scenario, space network access refers to the terminal can access the network and send message to any entity on the network at anytime and anywhere, while the user center can detect the terminal directly and exchange information with each other [6]. Therefore, we propose a scheme of constructing the space information telemetry and control network based on S-band multiple access relay satellite system. This method can meet the Interactive demand at anytime, anywhere and any node, and achieve system resources allocation optimization and system efficiency maximization. This method can change the traditional plan-driven control mode and realize the network-level intelligent telemetry and control mode based on situational awareness, as shown in Fig. 1.

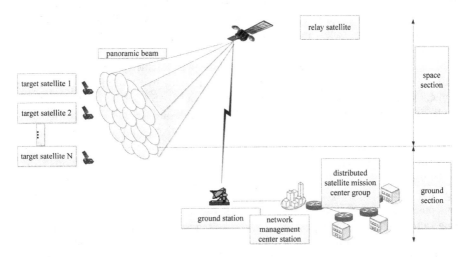

Fig. 1.   TT&C and management of SIN

In Fig. 1, the network architecture of the space information telemetry and control network based on S-band multiple access relay satellite system includes space section and ground section. The space section mainly includes relay satellites and target satellites installed with access terminals. The ground section mainly includes ground station of relay satellite, network management center station and distributed satellite mission center group.

The relay satellite system can achieve static full coverage of the Earth's surface and near-Earth orbit by building a panoramic beam through multi-satellite networking and SMA systems. Network management center station can send the real-time information returned from the target satellite to the user management center through the ground user network, while the user management center can send remote control command to the target satellite in real-time through the relay satellite system. The network management central station centrally manages the resources of all relay satellites. The user's resource request does not need to contain the information of which relay satellite to use and the network management center station can automatically allocate the most suitable channel resources according to the relay satellite position, the user platform position and the relay satellite channel usage. Figure 2 shows the resource application of the user of the SIN.

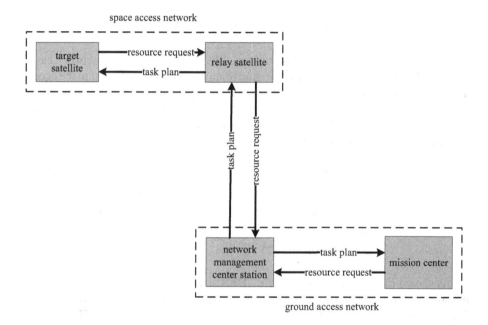

Fig. 2. User's resource request of SIN

As shown in Fig. 2, the user's resource application mode breaks through the traditional plan-driven control mode by constructing a space information telemetry and control network based on the SMA relay satellite system. User targets (with unique identity) under the SMA panorama beam coverage can initiate instant access applications and resource applications through the return channel (the system receives and updates the position coordinates and status information periodically reported by the user during the period when the user targets the network but does not communicate). Figure 3 shows the coverage of the panoramic beam of the space information telemetry and control network based on a SMA relay satellite system.

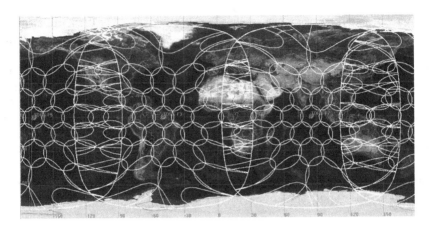

Fig. 3. Panoramic beam coverage

As shown in Fig. 3, the space information TT&C and management network based on the multi-satellite network and panoramic beam coverage is equivalent to the access network of the terrestrial cellular network. It can support information dynamic access of the users distributed around the world and the users from the surface or low altitude and the low orbit spacecraft users. Table 1 gives the analysis of the panoramic beam coverage of single TDRS satellite.

Table 1. The number of beam provided by a single TDRS satellite with panoramic beam

Coverage range	Beam number	Beam angle
Earth's surface	39	$\pm 8.5°$
200–1000 km	55	$\pm 10°$
200–2000 km	75	$\pm 11.5°$

As shown in Table 1, when a single TDRS satellite SMA system have 75 return beams (overlapping with 1.5 dB), it can achieve full coverage of the spacecraft below 2000 km and the whole combined beam angle is $\pm 11.5°$. In addition, the TDRS multi-satellite network can realize the static full coverage to the Earth's surface and near-Earth orbit. The users of the STN can access resources by random access and competition. Based on the global perception of user goals and on-line management of the whole network, the network management center can dynamically allocate the resources of the channel through the multi-dimensional arbitration of applications of the user platform and applications of the user center. This method can guarantee the traffic demand of large users and effectively reduces the time delay from the task establishment and the transmission to the user application.

4   The Preliminary Design of TT&C and Management for SIN

In order to implement the random access and online management of the users of the space information transmission system at any time and any place, it is necessary to rationally design the protocol mechanism, workflow and user terminal of the satellite-ground interactive command and information. According to the capability of SMA system of TDRS at present, the design of air interface is to solve the problems of random access, unexpected information transmission and application of online resources under the condition of large users. With the improvement of the space information transmission system capacity level, the design of the architecture for protocol can be upgraded to meet the integration of space and space information network application requirements.

4.1   Design of Architecture for Protocol

At present, data link layer protocols developed by CCSDS such as packet telemetry, subcontracting remote control, advanced in-orbit data systems, and near-space links, are mainly applicable to data systems of conventional spacecraft, large and manned spacecraft, the additional spatial links between the spacecraft close to each other [7]. The relaying of space information, mainly deal with the operation and processing of the physical layer and the data link layer. Because the relay satellite S-band multi-access system has the characteristics of long delay, asymmetric channel and high dynamic, and the management network involves multi-objective user's access, access control, channel allocation, resource application, monitoring and control network based on the relay satellite S-band multi-access system must have reliable, flexible and efficient communication protocols and process requirements for support. Therefore, the point is re-designing the physical layer and data link layer content based on the existing CCSDS protocol standards. The network layer, transport layer and application layer and other

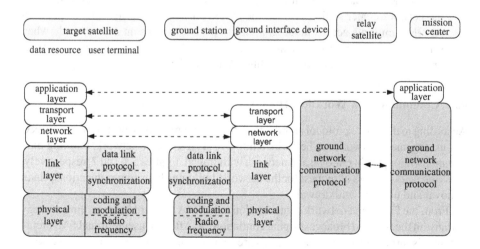

Fig. 4. Protocol architecture for TT&C and management of SIN

high-level protocol content can be designed according to system upgrades and functional development needs. Figure 4 shows the space information monitoring and management network protocol architecture based on the relay satellite S-band multi-access system.

Figure 4 shows that the air interface of the spatial information management and control network is mainly designed for the whole content and requirements of the physical layer and the relevant content and requirements of the data link layer, of which:

(1) According to the data transmission characteristics of relay satellite system and the application requirements of accessing burst data, the physical layer of space transmission link is divided into two sublayers, including RF sublayer and coding and modulation sublayer. RF sublayer mainly includes RF frequency, bandwidth, power, polarization and channel characteristics requirements. Coding and modulation sublayer includes coding, spread spectrum modulation, carrier modulation methods and performance requirements.

(2) The data link layer of the air interface is divided into two sublayers, including synchronization sublayer and the data link protocol sublayer. The synchronization sublayer is mainly used for the delimitation and synchronization of the transmission frame. The data link protocol sublayer is mainly used for framing data or data units from higher layers.

According to the above architecture for protocol, the forward and reverse channel resources of the S-band multiple access system are divided into a control channel and a plurality of traffic channels by CDMA code division:

(1) SMA forward broadcast channel is for all users, and broadcast channel contains the relay satellite's orbit position, resource usage and other information (using time division and code division multiple access mode, and time-sharing switching beam pointing).

(2) SMA reverse access channel for all users, the user platform can access the network to issue resources to use the application through the access channel. In the idle period, the user platform periodically reports its position and health information to the network center station through the access channel (using the spread spectrum ALOHA multiple access mode, a plurality of users shares an address code when sending the burst information, and if the number of users increases, may be appropriate to increase the number of public address code).

4.2 Communication Workflow Design

According to the above protocol architecture, the communication workflow of the TT&C and management network can be divided into five stages: access network, building link, communication, chain removal and network withdrawal. Figures 5, 6 and 7, respectively, gives the work process of request for the network, building chain communications, chain removal and network withdrawal.

From the Fig. 5, the network center station uses SMA forward beam polling strategy to transmit the forward broadcast signal continuously. After the user terminal SMA relay terminal is turned on, the forward signals are searched and captured. Then, the network

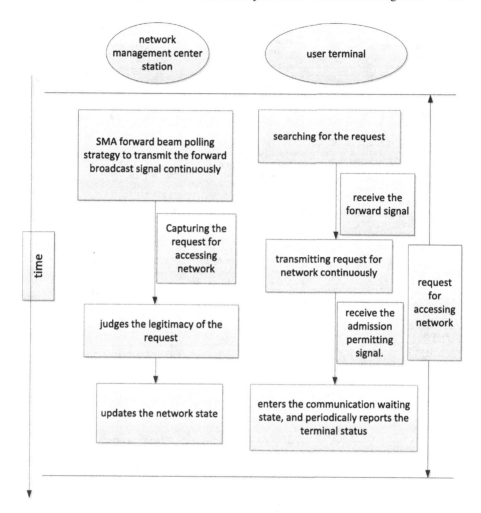

Fig. 5. Request for accessing network

admission request is transmitted continuously through the backward access channel. After the network center station receives the network admission request, it judges the legitimacy of the network admission request and sends the network admission permitting signal. Upon receiving the network admission permitting signal, the user terminal updates the network state, enters the communication waiting state, and periodically reports the terminal status.

From the Fig. 6, when the user platform has the spatial information transmission request, the resource use request can be sent through the SMA access channel. The network center station dynamically allocates the resources according to the resource usage and the user's service transmission demand, and sends the channel assignment message to the user through the SMA forward broadcast channel, and builds up

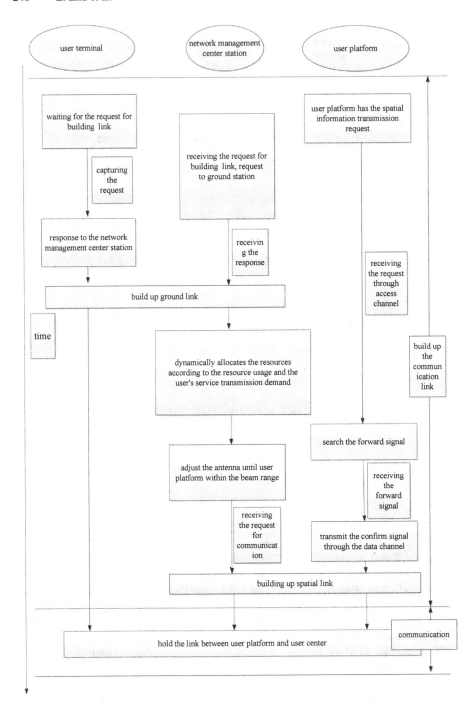

Fig. 6. Bulding up communication link

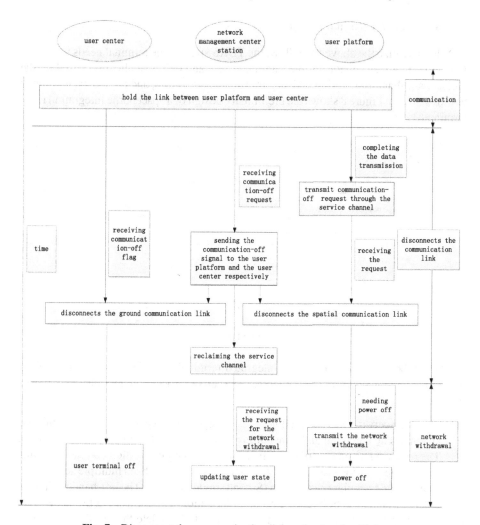

Fig. 7. Disconnect the communication link and network withdraw

spatial links and terrestrial links with the user platform and user centers to transmit the spatial information.

From the Fig. 7, when the user platform completes the data transmission, the communication-off application is sent through the service channel, after the network center station receives the application, it sends the communication-off signal to the user platform and the user center respectively, disconnects the communication link and reclaims the service channel; the user platform complete the network withdrawal by sending the withdrawal request and the network center station updates the user status according to the network withdrawal request.

4.3   Integrated Terminal Design

In order to realize the above work flow, the TDRS spaceborne terminal needs to add the functional modules supporting the satellite access and online application based on the traditional S-band monitoring and data transmission services, and realize the integrated terminal design. Figure 8 shows the preliminary design of a spaceborne integration relay terminal.

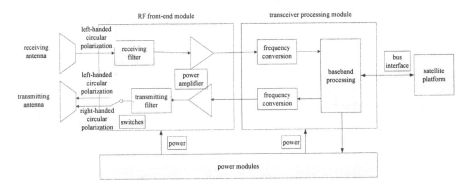

Fig. 8. Integrated terminal design

As can be seen from Fig. 8, the integrated terminal is made up by the transceiver antenna module, RF front-end module, transceiver processing module, and power modules.

(1) The transceiver antenna module uses independent receiving and transmitting antenna. The receiving and transmitting antennas are S-band wide beam antennas. The receiving antenna supports left-handed circular polarization and the transmitting antenna supports left-handed circular polarization (S-band multiple access) and right-handed circular polarization (S-band single-address). The left-handed circular polarization and right-handed circular polarization will not work at the same time.

(2) The main function of RF front-end module is to complete the pre-filter of the S-band forward signal and the amplification of the S-band backward signal, while using switches to switch the polarization mode.

(3) The main function of the transceiver processing module is to communicate (data flow, remote telemetry) with the satellite platform through the bus interface and at the same time to complete the conversion between the telemetry and remote control information to the forward and backward RF signals. The transceiver processing module comprises a transceiver frequency conversion channel and a baseband processing module.

In the access-free access mode, the integrated terminal adopts the pulse working mode. When the transceiver processing module sends the resource request information or the backward information, the control power module controls the power amplifier to power up; after the information is sent, the control amplifier is powered off to reduce the power consumption of the system.

5   Conclusion

According to the characteristics of spatial information network transmission demand and the development trend of the three big changes in the spatial information network user-controlled mode, the overall concept of space information monitoring and management network based on relay satellite S-band multiple access system is proposed. The system composition, panoramic beamforming, protocol architecture, communication workflow and so on are discussed and designed for the measurement and management for SIN. The system capacity, delay, data rate, multiple access interference and other technical indicators need to be demonstrated in the future, and the air interface protocol, data structure and other content need to be designed stratify. The research results of this paper break through the traditional plan-driven management mode, which could realize ubiquitous perception, random access, on-line management, and satellite-ground cooperation for users' application. Relevant result could be referenced in the study of measurement and management for SIN.

References

1. Jiasheng, W.: China's data relay satellite system and its application prospect. Spacecraft Eng. **22**(2), 1–6 (2013). (in Chinese)
2. Huiming, H., Baohua, K.: Architecture of space information transmission network using TDRSS as its backbone. J. Spacecraft TT&C Technol. **34**(5), 395–401 (2015). (in Chinese)
3. Baosheng, S., Shumin, G., Shuming, Y., et al.: Application modes of TDRS SMA system. J. Spacecraft TT&C Technol. **35**(1), 001–009 (2015). (in Chinese)
4. Goddard Space Flight Center. Space Network User's Guide (SNUG), 22 June 2015. http://esc.gsfc.nasa.gov/assets/files/450-SNUG.pdf
5. Xibin, S., Baosheng, S., Ligang, F., et al.: Study on operation modes of demand access service of TDRSS. J. Spacecraft TT&C Technol. **32**(2), 95–101 (2013). (in Chinese)
6. Rongjun, S.: Conception of earth-space integrated aerospace network of China. China Eng. Sci. **8**(10), 19–30 (2006). (in Chinese)
7. Weizhi, T.: Space Data System. Science and Technology of China Press, Beijing (2004)

Architecture and Application of SDN/NFV-enabled Space-Terrestrial Integrated Network

Xiangyue Huang$^{(\boxtimes)}$, Zhifeng Zhao, Xiangjun Meng, and Honggang Zhang

College of Information Science and Electronic Engineering,
Zhejiang University, Hangzhou 310000, China
{huangxiangyue,zhaozf,3120104519,honggangzhang}@zju.edu.cn
http://www.isee.zju.edu.cn/

Abstract. The space-terrestrial integrated network able to cope with the complex and varied tasks in future communications. However, there are some problems in unified management and cooperative scheduling. In this paper, based on the situation of satellite and terrestrial integrated network, combined with software defined network (SDN), network function virtualization (NFV), and mobile edge computing (MEC), we propose a new space-terrestrial integrated network architecture. SDN is used to implement centralized management and unified control. Distributed and unified virtualization platform was constructed based on the technology of NFV. The deployment of MEC over NFV platform helps to improve the user experience and network service quality. Finally, analyzed several typical scenarios such as supporting for multi virtual operators, improving the emergency response and coverage capabilities, optimizing service driven link and route under the proposed architecture.

Keywords: Space-Terrestrial integrated network · Software Defined Networking · Network Function Virtualization · Mobile Edge Computing

1 Introduction

Over the last few years, the spatial information network plays a key role in handling emergencies and accelerating the development of aerospace industry. With the development of spatial information network, combinating the satellite network and the terrestrial network to form an integrated communication network is becoming the trend to improve the communication services.

The space-terrestrial integrated network refers to the network composed of space-ed network and terrestrial-ed network, which are coordinated with each other and complement each other to realize integrated communication and service. In current satellite ground segment network architectures (referred simply to as satellite networks in the following), there is a lack of prevalent standards and much functionality is mainly deployed on vendor-specific network appliances. Although the integrated network is able to overcome the defects about

© Springer Nature Singapore Pte Ltd. 2017
Q. Yu (Ed.): SINC 2016, CCIS 688, pp. 244–255, 2017.
DOI: 10.1007/978-981-10-4403-8_21

the two networks, there are some new difficulties in the unified management of the space-terrestrial integrated network, and the problems of coordination and scheduling are still complicated [1,2].

Recently, SDN [3] has been introduced as a redesign of cellular network architecture that addresses the control and coverage issues, while reducing operational costs and improving heterogeneous multi-domain networks in agility, flexibility and scalability. Using the architecture of NFV [4] shows great potential in the services of multi-tenant and network autonomic management (self-optimization, self-adaptation, and self configuration), which could achieve flexible and efficient resource allocation and configuration. SDN and NFV in the integrated network will enhance the management capability and coordination level. Based on the virtualization common resource provided by NFV, MEC [5] are deployed at the relevant locations of the integrated Network to provide content, applications and computing services within the close proximity of users. Accordingly good for high-performance, content download and application implementation. Finally, improving user response speed, optimize network services and user experience.

Based on the development of the technology we mentioned before, we proposes a new architecture of integrated network integration based on SDN, NFV, MEC and other related technologies, to realize the space-based network and the terrestrial network complement each other in collaborative coverage and flexible control.

The paper is organized as follows: in Sect. 2, it begins by describing backgrounds of different technique in space-terrestrial integrated network. Section 3 provides an overview of the system architecture and function. In Sect. 4, we discuss several applications based on the architecture. Finally, Sect. 5 gives research conclusions and future directions.

2   Background

In this section, we will introduce the main technology in the Space-Terrestrial integrated network architecture, such as SDN, NFV and MEC. These techniques are being positioned as central technology enablers towards a improved and more flexible integration network.

2.1   SDN in Space-Terrestrial Integrated Network

SDN Concepts and Characteristics. SDN is a networking paradigm in which the forwarding hardware is decoupled from control decisions. It promises to simplify network management and utilize the centralized control plane and distributed forwarding plane. SDN technology has the following characteristics: the separation of control and data plane makes the network switches and routers into a simple forwarding device, the control logic is deployed in a centralized controller, network management platforms and applications use the controller to complete the software-based network configuration, monitoring, management, scheduling and optimization through the north interface. This makes the whole network highly programmable and configurable, which is conducive to deploy new strategies and new algorithms in control the granular of data flow precisely.

SDN Applications and Challenges. The advantages of SDN are centralized and dynamic control. In the integrated network, SDN can be used to implement centralized management and cooperative control of the whole network [6]. With the global state information of the integrated network, the requests of different users could be matched with the corresponding bandwidth and quality of service, and then choose the optimal path to achieve efficient use of the link resources. In addition, the SDN controller can be unified to manage and schedule of all network devices, to provide efficient and reliable link service.

SDN is being positioned as an enabler towards a new integrated network, but there are still many challenges [6]. First, from the SDN technology itself, the development of efficient cross-domain collaboration and inter-domain communication mechanism needs to be improved [7]. Secondly, the way to deploy SDN function into the satellite network ground station, the user gateway, remains unresolved; besides, the further research about unification and cooperation with the SDN function in the integrated network is needed.

2.2   NFV in Space-Terrestrial Integrated Network

NFV Concepts and Characteristics. NFV is a network architecture that uses the technologies of IT virtualization to virtualize entire classes of network node functions into building blocks that may connect, or chain together, to create communication services. So as to reduce the cost of dedicated communication network equipment (such as gateway, mobile switching center, firewall, etc.), and improve system flexibility and reliability. NFV makes network equipment function no longer dependent on dedicated hardware, resources can be fully flexible sharing, new network functions and new services can be quickly developed based on actual business needs. It also can achieve in flexible service, fault isolation and rapid self-healing and so on. In addition, dynamic slicing and functional combination of communication networks can be realized based on NFV [8], in order to realize the support for different services, various virtual private networks and multi-tenants, and meet the flexible operation needs of the future network.

NFV Applications and Challenge. In Space-Terrestrial integrated network, utilizing the NFV technology to build a integration of distributed network virtualization platform is an essential fundamental work, all the various required types of network equipment could be build through the standardization of the interface [9]. The dynamic resource scheduling function of NFV provides a high degree of reliability and self-healing features, to ensure reliable operation of network equipment and emergency services. In addition, through the NFV-based network dynamic fragmentation, it can provide different types of business and users with personalized virtual network.

Applying NFV in Space-Terrestrial integrated network is clearly that much work should be done. Firstly need to develop and implement the related network equipment based on NFV, secondly to solve the problem of cooperation and smooth evolution between the traditional network equipment and the new equipment based on NFV. In addition, different requested virtual network function (VNF) units are planned on the top floor.

2.3   MEC in Space-Terrestrial Integrated Network

MEC Concepts and Characteristics. MEC refers to provide storage resources and cloud-computing capabilities at the edge of the mobile network (base stations, terminals) in close proximity to users. Which accelerates the download of content and promotes the local service such as subscriber location that can be used by applications and services to offer context-related services; Even more, it offer feedback services so that service are fully utilized through adapt to wireless network conditions dynamically [10]. For application developers and content providers, MEC will provide open platform and unified programming interfaces, making the mobile network capabilities and status open to the applications directly. This can not only improve the efficiency of applications and services, save transmission bandwidth and reduce the delay, but also change the communication "pipeline" into a "capacity platform", and make full use of base stations, terminals and other computing and storage capabilities to achieve efficient services.

MEC Applications and Challenge. The bandwith resources are more valuable in the Space-Terrestrial integrated network, we deploy MEC in the satellite network gateway, satellite user gateway, base station, user gateway and so on to take advantage of the resources. All this will help make full use of localization Calculation and reduce the bandwidth for wireless channels, satellite channels, backhaul network, improve communication efficiency finally.

It is a new topic to apply MEC architecture in the Space-Terrestrial integrated network. The first part of applying MEC is primarily about sorting out the functions (local computing and storage, location information, channel information, terminal information, etc.) that the MEC can provide, which the Space-Terrestrial integrated network needed according to the application and service type provided. Then edge computing and storage capabilities should deploy at appropriate locations to achieve goals in the aspect of economic, efficient and reliable.

3   System Architecture and Function

We proposed a SDN/NFV-enabled space-terrestrial integrated network architecture, as shown in Fig. 1. The architecture consists of user gateway, terrestrial base station, satellite ground gateway and so on, SDN and NFV play a great role in each component.

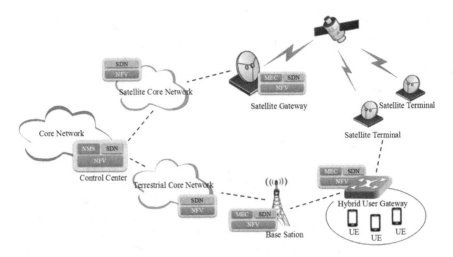

Fig. 1. System architecture.

3.1   System Functional Unit

User Gateway. The user gateway requires the ability to access the satellite network and terrestrial network, which performs more flexible in connection for users. Implementing the SDN technique in the user gateway that helps forward traffic according to the rules of flow table issued by the controller, therefore different traffic can be distinguished at the user gateway before transmitted. To manage and utilize the hardware efficiently in user gateway, NFV technique is in a position to virtualize its hardware resources, give its workloads great flexibility to achieve firewall, deep packet inspection and other network functions. The MEC function is deployed on the foundation of NFV, which allows the compute-intensive application and high bandwidth occupied content be processed directly at the edge of network, This is going to reduce the network delay and ease the burden of data center. MEC-based content caching can improve the efficiency of content distribution and enhance the user experience.

Terrestrial Base Station. The terrestrial base station also utilizes the structure of NFV structure as well [11]. Base station at the NFV, on the one hand can achieve the base station function virtualization and functional fragmentation to meet the future broadband, car networking, Internet of things and other business needs, on the other hand can also support multi-operator virtualization, improve Resource utilization, reduce power consumption. They deploy MEC at the base station based on NFV, you can complete design different strategy in content distribution, providing localization services, location-based and delay-sensitive services.

Satellite Gateway. The satellite gateway is applied to unify the information of the satellite stations into the satellite backbone network. The SDN controller here is to control all the SDN switches in satellite stations. Beyond that, SDN switch is another equipment to distinguish and forward traffic based on the flow table. NFV can realize the virtualization of the gateway hardware resources, and realize the VNF function such as the gateway function and the SDN controller function through the software programming. MEC based on NFV can implement content caching and location-based content services.

Backbone Network. The SDN controllers in the satellite terrestrial backbone networks and terrestrial backbone networks are responsible for controlling the SDN forwarding devices in respective domains, such as satellite gateways and base stations. NFV can realize the virtual network resource, therefor various network functional units in satellite networks and terrestrial wireless networks are instantiated such as switching center, position register, authentication center and so on.

Core Network Service Control Center. The core network service control center is a cross-domain coordination and centralized control center of the Space-Terrestrial integrated network, which connects the satellite ground backbone network and the wireless ground backbone network. The SDN controller in core network service control center is the top-level controller, which has the ability to manage the lower level SDN controller in satellite and terrestrial domains. NMS (Network Management System) as a part of the control center calls north interface to complete the software-based network configuration, monitoring, management, scheduling, optimization and so on [12]. It can also complete the dynamic configuration and partition of NFV, and manage the resource requirements based on user requests and hot events.

3.2   Typical Internal Structure

Satellite Ground Gateway. The foundation of the satellite ground gateway is supported by NFV, As shown in Fig. 2, all the virtual network functions (including SDN controller, virtual router, firewall, MEC-Cache and MEC-LBS) is implemented on it. The NFV management module which has the functions of resource management and allocation is in charge of instantiating the VNF. At the top of the layer are the various function entities provide the service. The SDN controller is responsible for controlling the switch. The virtual router is used for the packet forwarding and filtering according to the flow table rules issued by the upper controller. The MEC-Cache component provides caching services; MEC-LBS provides services based on "location" (i.e., the terrestrial gateway-related).

Satellite User Gateway. As shown in Fig. 3, Satellite user Gateway is similar with the Ground Gateway, the VNF (including the SDN switch, Firewall, MEC-Cache, MEC-Encoder) required by users is implemented in software programing

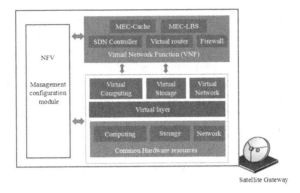

Fig. 2. Internal function architecture of satellite ground gateway.

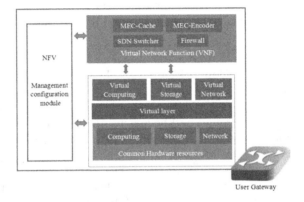

Fig. 3. Satellite user gateway functional architecture.

on the common computing, storage and network hardware resources through the management module of NFV itself. Where the SDN switch is controlled by the superior SDN controller to achieve flow classification and forwarding; Firewall is used for packet monitoring and filtering; MEC-Cache is going to provide caching services (content, software, etc.); MEC-Encoder is responsible for providing local video compression services before uploading.

4 Application Scenarios

4.1 Virtualization and Multi-tenancy in Integrated Network

The deployment of NFV in user gateway, ground base stations, satellite gateways, backbone networks and core network makes the whole network platform to achieve the virtualization. As a virtualized service (NFVIaaS) [13], the infrastructure of the entire Space-Terrestrial Integrated network can provide different virtual operators with a "standalone" operating environment, enabling the entire network infrastructure to support multi-virtual operator applications, as well as reunification for military and civilian in the same framework.

As shown in Fig. 4, virtual operators 1 and 2 run their own virtual network functions on the infrastructure provided by NFV architecture, which provides service resources for different operators matching with their business services. The virtual operator possesses a complete independent and self-management network in its own view.

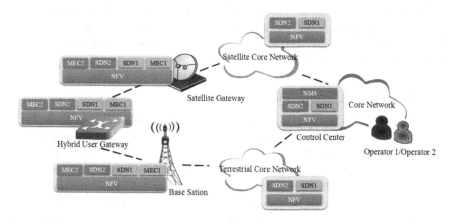

Fig. 4. Virtualization and multi-tenancy in integrated network.

4.2 Cooperative Coverage Increases Emergency Capability

As shown in Fig. 5, the SDN/NFV-enabled space-terrestrial integrated Network architecture achieves co-coverage on maritime, aviation, and land. Satellite networks and terrestrial networks are achieved to full coverage in a collaborative

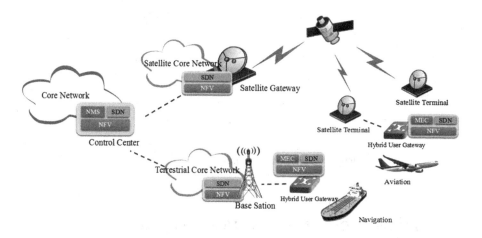

Fig. 5. Cooperative coverage increases emergency capability.

approach. For example, in some remote mountainous areas, or the area that the base station is not easy to build, the use of satellite networks to cover is an economic and practical choice [14].

In this scenario, The integrated architecture not only makes the coverage more widely expended user access, also makes the network be more stable and robust in case of emergency or hot issues on account of more paths to choose from. When the terrestrial base station is damaged and unable to communicate, the SDN controller can switch and configure the link flexibly by changing the flow table forwarding rules. The support of NFV enables the flexible scheduling and allocation of resources, which can provide resources flexible matching under different situations [15], and overcome the pressure of a large number of communication service demand in a short time.

4.3   Link Optimization Based on Service

A service-oriented link optimization approach could allow the communication resource full used. As shown in Fig. 6, based on the deep packet inspection function of NFV, the service flow can be classified at the user gateway. Leveraging the function of SDN controller in dynamic flow table control, different links can be planned for different priority services, the content of lower requirements on transmission delay can be transmitted through the satellite network, otherwise, replaced by ground based transmission. All this benefits user experience and link utilization. In the SDN unified control and scheduling, the separation of the uplink and downlink achieved support more flexible applications and services [16].

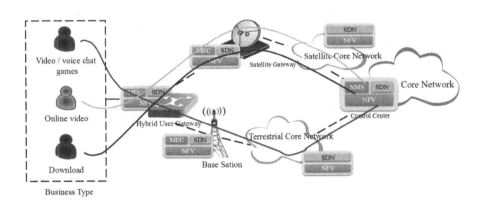

Fig. 6. Link optimization based on service.

4.4   MEC-Based Data Fusion and Upload in Edge

As shown in Fig. 7, the MEC in the user gateway can realize the initial fusion of the uploaded content with large data volume (video, Internet of things, etc.).

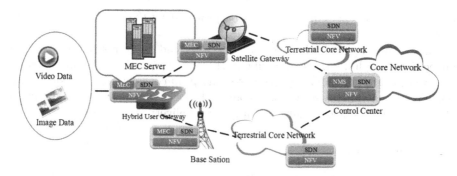

Fig. 7. MEC-based data fusion and upload in edge.

The way of fusion such as video compression, key frame drawing and compression of I/O sensor redundant data, Through the local computing power, reduce the amount of data uploaded, to improve the efficiency of communication and reduce communication delay [17].

This approach not only reduces the network latency, but also can transfer the computational burden of the cloud computing center to the edge nodes, and realize the cooperative calculation of the center end and the edge of network.

4.5   MEC - Based Edge Content Caching and Multicasting

For the content distribution application, the MEC storage resource is used to cache frequently accessed and popular content mainly in reducing the delay of the content requested. The user gateway or terrestrial base station connected to the terrestrial satellite station implements cooperative caching, and when the

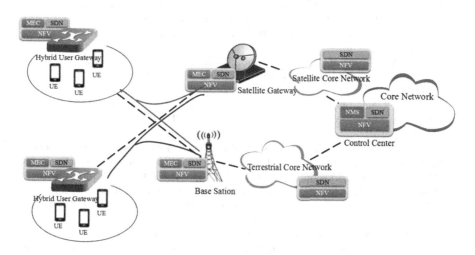

Fig. 8. MEC-based edge content caching and multicasting.

user initiates content requests, it can be provided by the MEC caching server directly if the content cached.

As shown in Fig. 8, both the satellite network and the terrestrial base station are equipped with multicast capability. MEC cache is used at the satellite gateway and the ground radio network base station, in consideration of requests for popular content in this scenario, it could achieve more efficient in distribution of content by considering multicast in content delivery [18].

5   Conclusion

Considering the development of satellite and terrestrial communication network, this paper proposes a new Space-Terrestrial Integrated network architecture based on SDN, NFV, MEC and other related technologies to realize the advantages of Integrated network. This paper analyzes the effective support on multi-operator services, efficient collaborative coverage, service-driven routing, data fusion and uploading, content cache multicast and other application scenarios, the flexibility and effectiveness of the new architecture are briefly analyzed.

The next step will focus on the virtual network function under NFV support, multi-layer SDN architecture, MEC-based application design and performance evaluation, to achieve more efficient, reliable and flexible service.

References

1. Zhang, N., Zhao, K., Liu, G.: Reflections on the construction of "information network of integration of heaven and earth" in China. J. China Inst. Electron. Sci. **10**(3), 223–230 (2015)
2. Shen, R.: China's integration of space and space of the concept of interconnection network. China Eng. Sci. **8**(10), 19–30 (2006)
3. Kreutz, D., Ramos, F.M.V., Verssimo, P.E., Rothenberg, C.E., Azodolmolky, S., Uhlig, S.: Software-defined networking: a comprehensive survey 103, 1–76 (2015)
4. ETSI, NFV G S. 002-2013: Network Functions Virtualization (NFV). ETSI Press, France (2014)
5. Ahmed, A., Ahmed, E.: A survey on mobile edge computing. In: IEEE International Conference on Intelligent System and Control, pp. 1–8. IEEE Press, Tamil Nadu, India (2016)
6. Ferrs, R., Koumaras, H., Sallent, O., et al.: SDN/NFV-enabled satellite communications networks: opportunities, scenarios and challenges. Phys. Commun. **18**(2), 95–112 (2016)
7. Zhang, W., Hong, P., Yao, L., et al.: CSRS: a cross-domain source routing scheme for multi-domain software-defined networks. In: IEEE International Conference on Communication Workshop (ICCW), pp. 375–380. IEEE Press, London (2015)
8. Zhou, X., Li, R., Chen, T., et al.: Network slicing as a service: enabling enterprises' own software-defined cellular networks. IEEE Commun. Mag. **7**(54), 146–153 (2016)
9. King, D., Farrel, A., Georgalas, N.: The role of SDN and NFV for flexible optical networks: current status, challenges and opportunities. In: International Conference on Transparent Optical Networks (ICTON), pp. 1–6. IEEE Press, Budapest (2015)

10. Cau, E., Corici, M., Bellavista, P., et al.: Efficient exploitation of mobile edge computing for virtualized 5G in EPC architectures. In: IEEE International Conference on Mobile Cloud Computing, Services, and Engineering, pp. 100–109. IEEE Press, Oxford (2016)
11. Trivisonno, R., Guerzoni, R., Vaishnavi, I., et al.: Network resource management and QoS in SDN-enabled 5G systems. In: IEEE Global Communications Conference (GLOBECOM), pp. 1–7. IEEE Press, San Diego (2015)
12. Zhou, J., Jiang, H., Wu, J., et al.: SDN-based application framework for wireless sensor and actor networks. IEEE Access **4**(3), 1583–1594 (2016)
13. Giuseppe, C., Luca, F., Pi, A., et al.: Quality audit and resource brokering for network functions virtualization (NFV). In: Orchestration in Hybrid Clouds, pp. 1–6. IEEE Press, San Diego (2015)
14. Iapichino, G., Bonnet, C., Herrero, O., et al.: Advanced hybrid satellite and terrestrial system architecture for emergency mobile communications. In: International Communications Satellite Systems Conference (ICSSC), pp. 1–8. AIAA Conference Pubilcation, San Diego (2008)
15. Jean, S., Cho, K.: The structure for distributed virtual resource control to provide agile network service in NFV. In: Information and Communication Technology Convergence (ICTC), pp. 826–829. IEEE Press, Jeju Island (2015)
16. Xu, H., Chang, Y., Sun, T., et al.: Modeling and analysis of uplink-downlink relationship in heterogeneous cellular network. In: International Symposium on Personal, Indoor, and Mobile Radio Communication (PIMRC), pp. 1338–1342. IEEE Press, Washington DC, (2014)
17. Nunna S., Kousaridas A., Ibrahim, et al.: Enabling real-time context-aware collaboration through 5G and mobile edge computing. In: Information Technology-New Generations (ITNG), pp. 601–605. IEEE Press, Las Vegas (2015)
18. Poularakis, K., Iosifidis, G., Sourlas, V.: Exploiting caching and multicast for 5G wireless networks. IEEE Trans. Wireless Commun. **15**(4), 2995–3007 (2016)

Service Customized Software-Defined Airborne Information Networks

Xiang Wang$^{(\boxtimes)}$, Shanghong Zhao, Jing Zhao, Yongjun Li,
Haiyan Zhao, and Yong Jiang

School of Information and Navigation, Air Force Engineering University, Xi'an, China
lleafwx626@126.com, wangxiang_626@hotmail.com

Abstract. Airborne information networks have drawn the attention of the research community in the recent years. This growing interest can largely be attributed to new civil and military applications. Currently Airborne information networks are inherently hardware-based and rely on closed and inflexible architectural design. This imposes significant challenges into adopting new communications and networking technologies. Software Defined Networking, recognized as the next-generation networking paradigm, relies on the highly flexible, programmable, and virtualizable network architecture to improve network resource utilization, simplify network management, and promote innovation and evolution. In this paper, the concept of Software Defined Networking technology is introduced first. The features for the airborne environment and research challenges to realize the next-generation airborne information networks are discussed. Furthermore, a software-defined architecture is proposed to facilitate the development of the next-generation airborne information networks. The function of network virtualization was added, the SDN airborne information network provide differentiated services for airborne applications through employing the network virtualization technology. Aiming at the characteristics of airborne information network environment, a series of key technologies are analyzed in detail.

Keywords: Airborne information network · Software Defined Network

1 Introduction

Airborne information networks, which are widely employed in emergency relief, transoceanic navigation, environmental science, communication relay and so on, are defined as a network composed of high-speed moving plane as the primary node and wireless communication link between the nodes [1, 2]. With the development of wireless communication technology, network transmission ability, network scale and network applications diversity, the future airborne information networks will be IP-based network while facing numerous problems and challenges at the same time, such as high mobility, large span and spare distribution, the instability in link quality large accessing user size, the diversification of the aeronautical communication demand [3, 4]. However, for the sake of those problems the most present research works always have been based on TCP/IP protocol architecture and integrated new functionality in network nodes. Generally,

© Springer Nature Singapore Pte Ltd. 2017
Q. Yu (Ed.): SINC 2016, CCIS 688, pp. 256–265, 2017.
DOI: 10.1007/978-981-10-4403-8_22

multiple protocols should be work together to satisfy these network requirements, which leads to disadvantage influence on the high complexity of the design and implementation and the low efficiency in processing information.

Software defined network (SDN) provide an opportunity to deal with the aforementioned problems. As a novel network architecture, SDN clearly separates the data plane from the control plane in a network. The former is run by network switches to forward packets according to the logics set by the latter through a standard API. Accordingly, the control plane is responsible for network configuration and management which are set by a software program [5–7]. In addition to cope with a series of adaptive problems which brought by the new generation airborne information networks based on SDN technology, the certain application demand with diversified quality-of-service (QoS) is a uprising challenges should be faced. Because of QoS differentiation for airborne communications in airborne networks, network resources, for example compute, storage and bandwidth, could be configured effectively by the characteristics of various information. Through constructing virtual network with DiffServ QoS, network service can be customized for different operation and application, thereby enabling differentiated services, flexibility transmission mode and more efficient utilization of network resource.

In order to solve the adaptability issues of the aeronautical environment and to meet the different needs of different application of the airborne services, software-defined airborne information networks with the ability to customized service are is proposed in this paper. The rest of this paper is organized as follows. The principle and connotation of SDN is described in Sect. 2. The technology challenges in the airborne information networks are described in Sect. 3. Furthermore, the network architecture of software-defined airborne information networks is proposed to facilitate the development of the next-generation airborne information networks in Sect. 4. Section 5 provides a series of key problems and solutions to the application of airborne customization network in the airborne environment. In the end, Sect. 6 presents conclusions of our study.

2   The Principle of SDN

As a new network structure, the centralized control plane and the distributed data plane are employed in SDN and the two planes are separated. The control plane uses the control-distribute interface to centrally control the network devices on the data plane and supports a flexible programmable ability [8, 9]. The basic structure is shown in Fig. 1.

The designing structure of SDN mainly includes three layers of the data transmitting layer, controlling layer, information layer and the interfaces between adjacent planes:

(1)   The data plane that consists of bottom transmitting devices major in processing the data based on operation flow table, transmitting and status collecting. In addition, according to the command that issued by the control plane to quickly transmit the data.

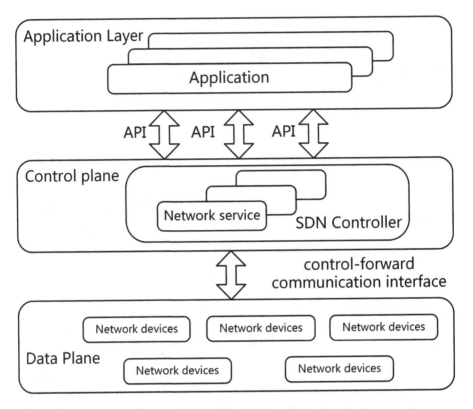

Fig. 1. Conceptual sketching of software defined networks.

(2) The control plane mainly processes the abstract information that come from the source on the data plane. On one hand, the control plane can control the bottom network devices to carry transmitting task and abstracting the sources from the bottom network devices to achieve the information. After that, an abstract view of the networks can be generated.

(3) The application plane mainly includes the applications that aim to different network services in the control plane. It majors in the methods of transmitting and processing the applications by managing and controlling the network.

Thus, comparing to the traditional network structure, there are three main advantages for the network structures of SDN [10, 11]:

(1) The logical centralized control can be realized based on the network control (achieved by software) and data exchanging (achieved by hardware).

(2) The networks is programmable. It can support different QoS methods for different flow tables and different operations which expands the flexibility and the multi-service ability.

(3) The centralized network control. The whole control view can be achieved from a single controller through control software to make a decision of transmitting actions based on the overall information. It can avoid the network congestion. In the

meantime, choosing a most proper transmitting tracking for the operation and application demands.

3 The Technology Challenges in the Airborne Information Networks

The airborne information networks regarded the high speed planes as the main nodes, including some other motivation objects in the air. It is a network that consists of a laser communication links, which is regarded as a backbone structure, and the wireless communication links among the nodes to transmit packets and support IP service. The airborne information networks possess the properties of large covering range, high dynamic condition and high rate, dynamic self-organization ability, etc. It is an important method to guarantee the connectivity among the nodes and support communication service under the condition of the limit communication fundamental during emergency rescue, target detecting and tracking, communication relay, coordinated combat. What's more, it can also connect satellite networks and the ground networks to form an air-space-ground cooperated distributing network which has a good prospect.

(1) High mobility, large span and spare distribution. The plane nodes distribute sparsely in the large scale air scene. Both the high flying nodes and the low flying node are simultaneously existed. The differences of time delay among the planes are obvious. The plane nodes usually move quickly relative to the ground, the speeds among the nodes also move very quickly and the topology of the networks changed quickly [1]. A problem of connection in airborne information network will be brought by the distribution of high dynamic and large scale nodes. The airborne information network plays an important role in the operation of emergency rescue, target detecting and tracking, communication relay, etc. It needs to make sure that the destroyed nodes will not influence other nodes to remain a good performance and the properties of the network can be guaranteed.

(2) Link instability. The backbone network in the airborne information network employed the laser links to share the information, and the sub-network use the microwave links to carrier line-to-sight communication. In the air to air, air to ground and air to space links, the laser links suffered greatly by the air channel. The absorbing and scattering in the air will cause power fading that influence the optical receiver. The fluctuation of light intensity and atmospheric shimmer that caused by atmospheric turbulence impact the error rate and bit rate [12]. The microwave links will be influenced by the rains, snows, foggy etc. and route fading, multipath effect [13]. Thus, the airborne links will interrupt and rebuild frequently, suffer high error rate and large time delay. These are the challenges for the connectivity and feasibility in the airborne information networks. In order to raise the feasibility and tolerance, the reliability should be well considered when deploy multi-controllers. To guarantee the properties, extend and robust through choosing the number and locations of controllers.

(3) Diversified QoS. On one hand, the main operation of airborne information networks including emergency rescue, target detecting and tracking, communication relay,

coordinated combat [14–16]. For the target spy of tracking, a low time delay and high reliability is desired. For the statue information, the link capability is demanded highly. And the time delay is often desired between 10 ms and 100 ms. As for the remote sensing relay, the low jitter and high bandwidth are necessary. On the other hand, in order to satisfy the diversified demands on the upper layer, a logical isolation virtual sub-network is desired to build. Meanwhile, different decisions can be adopted according to self-requirement demand in the networks.

4 Architecture of Service Customized Software-Defined Airborne Information Networks

Software-defined airborne information networks are designed based on the idea of software defined networks to meet with the demands of future airborne information networks applications. It is a new generation of airborne information networks with open extensible and flexible information scheduling, which make it possible to provide differentiation service for future airborne information networks. The main characteristic of this network is that by establishing a virtual network with different Qos levels, and this network allows airborne applications with different levels choose their proper service level based on demands. As a result, this network can provide differentiation service to the whole network and enhance communication efficiency significantly. The conception of software-defined airborne information networks is proposed in this paper, and the architecture of service customized SDN airborne information networks is designed. Sketch map of the architecture is given in Fig. 2.

It can be seen that software-defined airborne information networks mainly consist of three parts:

(1) Mobile sub-nodes, including air access nodes and ground mobile nodes, can collect and transmit multiple state data through shortwave, ultrashort wave and laser wireless communication links. Air access nodes consist of high speed and high maneuverability manned and unmanned aerial vehicles. Ground mobile nodes include ground base, mobile tactical nodes, mobile information terminals and information processing centers.

(2) Airborne backbone nodes, consisting of large aircraft platforms and unmanned reconnaissance vehicles with high stability, can be employed as data receivers to receive and transmit state information from mobile nodes. Besides, airborne backbone nodes can also be used as network controller to manage and control the network.

(3) In-band and Out-band control channels allows the transmission and exchange of control information among mobile nodes and controllers. In-band mode means the control information is transmitted using the same channel with data. Out-band mode means the control information is transmitted using a dedicated transmission channel, namely control information and data are transmitted through different channels. Laser communication links with high bandwidth and low time delay are utilized as in-band control channel for multi-hop conditions, while microwave links

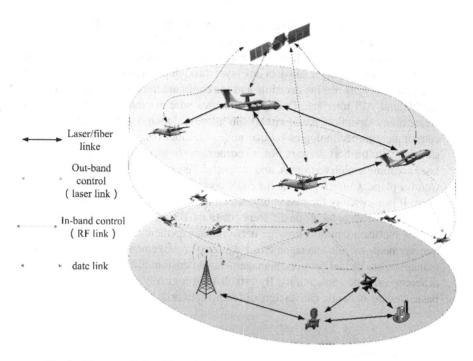

Fig. 2. Conceptual sketching of software-defined airborne information networks.

with low bandwidth and long distance are utilized as out-band control channel for single hop conditions.

Different from traditional airborne system based on wireless microwave technology, the service customized soft-defined airborne information networks have multiple hardware platforms loaded on each sub-node, including short-wave/ultrashort-wave radios, airborne radio frequency antennas and wireless laser communication terminals, which can support different types of communications. The form of software programming can be used in mobile sub-nodes in the networks, by using unified and open control interface, such as Openflow and SNMP technology, the nodes can realize the functions of physical layer, data link layer and network layer, and subsequently obtain the ability of network function virtualization (NFV). The NFV can enhance the flexibility of network structure, and in this way, consumers can use communication technology and network project flexibly with simple software arithmetic. In addition soft-defined airborne information networks are able to abstract the underlying physical network and custom virtual network according to different network demands. Then shared reuse of network basic equipment with different service applications is achieved, namely the network function virtualization. In accordance with the above ideas, system structure of soft-defined airborne information networks is shown in Fig. 3. The system consists of data plane, control plane, application plane, northbound interface and southbound interface.

(1) Data plane. Data plane consists of open, programmable and virtualized network transmission nodes, including airborne backbone nodes, air access nodes, ground

mobile devices and ground base. Specific network function module is depicted in Fig. 3. Data plane provide software radio module, Openflow flow, wireless management program and multiple hardware platforms. Software radio module realizes the function of physical layer and data link layer through programming. Openflow flow can define network routing decision, and can configure the network controller using southbound API interface. Data platform uses wireless management program to create many Openflow based virtual sub-node, and then realize different wireless communication technologies on one node. Multiple hardware platforms include short-wave/ultrashort-wave radios, airborne radio frequency antennas and wireless laser communication terminals, and support different types of communications.

(2) Control plane. Control plane is the SDN controller located on airborne backbone layer. It is a logically focused entity, which means all the controllers can locate at the same position or locate on different positions. Besides data forwarding function, control platform can also provide network management, including routing, flow management, mobile management, frequency management and virtual network management. Virtual network management can custom differentiated services for different airborne applications. By establishing logically isolated virtual subnets, basic device network can assign different network, bandwidth and computing

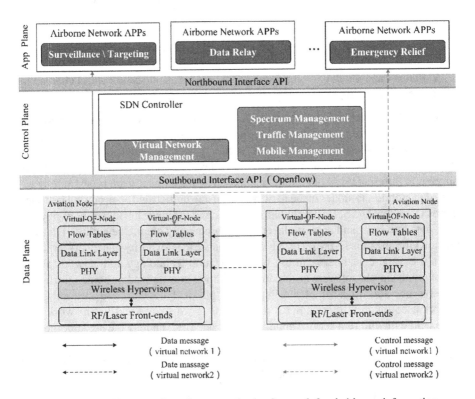

Fig. 3. Network architecture of service customized software-defined airborne information networks

resource to airborne applications based on different application demands. Meanwhile, airborne businesses can also custom differentiated network transmission strategies according to specific demands. Subnets are virtually isolated to ensure the security of data transmission.

(3) Application plane. Application plane is made up of many airborne SDN applications, and it is the most attention application for airborne network consumers. Airborne SDN applications usually contain emergency rescue, target detecting and tracking, communication relay, coordinated combat. These applications can submission their network behavior requirements to controller by programmable methods, and then accomplish the interaction with SDN controller.

(4) Northbound interface and southbound interface. Northbound interface is a series of interfaces among application plane and control plane. It mainly provide abstract network view, which enables airborne SDN applications to control the behavior of networks. Southbound interface mainly supports Openflow protocol, and the specific versions are V1.0 and V1.3. Southbound interface enables network management programs to access and manage the resources and devices in networks.

5 The Key Techniques in Service Customized Soft-Defined Airborne Network

5.1 System Structure with High Robust

This technique is mainly employed to guarantee the robust of the service customized soft-defined airborne network. The traditional SDN technique utilizing the centralized control method, however, it hard can be employed in the high dynamic, large scale sparseness distributed airborne information networks, because that if only one controller is broken, the whole network cannot properly work. At the same time, the network scale is increased along with the increase of the size of the network, however, the process and expansion capabilities capability of a single controller is very limited, which may hinder the applications. Therefore, multiple controllers with logical centralized and physical contributed structure should be considered for the network, to reducing the load of the controller and improving the comprehensive performance of the network.

5.2 The Controllers Deployment

Deploying multiple controllers with contributed structure can improve the expansion and robust capabilities, as well as the network performance. When the number of the controllers is set, the local deployment of the controller and the selection of the network nodes can be cataloged as NP-hard problem, in which the performance parameters such as the transmission time delay, network robust and controller load proportion should be considered and optimized. In large scale three dimension airspace scene, making the airborne backbone nodes as the controllers will cause a moderate or high time delay in the network. To cope with this problem, the controllers should be deployed in both backbone and child nodes. Therefore, how to realize the optimal deployment and distribution of the controllers is a key problem.

5.3 The Service Customized Technique

This technique is mainly guarantee the efficiency utilization of the network resources and equipments for different service and applications in airborne environment. Airborne information network has different services and applications, meaning that different services should be provided according to the specific application demands. By constructing logically isolated sub-networks, the network, bandwidth and link resource can be customized corresponding to the requirement of the application, to realize the high effect and flexible utilize of the network resource, as well as achieve the network virtualization. Therefore, how to construct the virtualized network and realize the mapping and isolation functions of the virtualized network is another key problem.

5.4 The Outage Recovery Mechanism

In high dynamic, large scale airborne environment, the airborne nodes are very sensitive to the environment variations, and the outage of the airborne nodes will affect the network performance. The fault in the SDN network can be classed as two kinds, i.e. controller error and network error. The network error consists of node and link errors. Specifically, the error of a single node may induce lapse of many links, reduce the connectivity of more nodes in the network, and damage the network performance. Therefore, the outage recovery mechanism should be established, in which both the initiative protection technique and passive reconstruct should be considered.

6 Conclusion

The future airborne information network will be an evolution based on full IP technique. The improvement of the wireless communication technique and network transmission ability, and the diversification of the application service are taking great changelings to the traditional IP network. The SDN technique is a promising method to overcome these problems, however, a series of airborne environment adaptability problems is emerged. This paper is worked on the application of the SDN technique in airborne information network. The principle and essential of the SDN technique is introduced, the key problems associated the application of the SDN technique in airborne environment is analyzed, and the basic structure of airborne information network based on SDN technique is proposed, finally, the key problems of the airborne customized network under airborne communication environment is investigated.

References

1. Kwak, K., Sagduyu, Y., Yackoski, J., Azimi-Sadjadi, B., Namazi, A., Deng, J., Li, J.: Airborne network evaluation: Challenges and high fidelity emulation solution. IEEE Commun. Mag. **52**(10), 30–36 (2014)
2. Sakhaee, E., Jamalipour, A., Kato, N.: Aeronautical ad hoc networks. In: Proceedings of the IEEE Wireless Communications and Networking Conference (WCNC), Las Vegas, pp. 246–251. IEEE (2006)
3. Nielsen, T.T., Oppenhaeuser, G.: In orbit test result of an operational optical intersatellite link between ARTEMIS and SPOT4, SILEX. In: SPIE, vol. 4635, pp. 1–15 (2002)
4. Fidler, F., Knapek, M., Horwath, J., et al.: Optical communication for high-altitude platforms. IEEE J. Sel. Top. Quantum Electron. **16**(5), 1058–1070 (2010)
5. Akyildiz, I.F., Lin, S.C., Wang, P.: Wireless software-defined networks (W-SDNs) and network function virtualization (NFV) for 5Gcellular systems: An overview and qualitative evaluation. Comput. Netw. J. **93**(1053), 66–79 (2015)
6. Akyildiz, I.F., Wang, P., Lin, S.C.: SoftAir: A software defined networking architecture for 5G wireless systems. Comput. Netw. J. **85**, 1–18 (2015)
7. Hong, C.-Y., Kandula, S., Mahajan, R., Zhang, M., Gill, V., Nanduri, M.: Achieving high utilization with software-driven wan. In: Proceedings of the ACM SIGCOMM, pp. 15–26 (2013)
8. Feamster, N., Rexford, J., Zegura, E.: The road to SDN: An intellectual history of programmable networks. ACM SIGCOMM Comput. Commun. Rev. **44**(2), 87–98 (2014)
9. Nadeau, T.D., Gray, K.: SDN Software Defined Networks. O'Reilly Media, Inc., Sebastopol (2013)
10. Mckeown, N., Anderson, T., Balakrishnan, H., et al.: Open Flow: Enabling innovation in campus networks. ACM SIGCOMM Comput. Commun. Rev. **38**(2), 69–74 (2008)
11. Ali, S.T., Sivaraman, V., Radford, A., Jha, S.: A survey of securing networks using software defined networking. IEEE Trans. Reliab. **64**(3), 1086–1097 (2015)
12. Sodnik, Z., Lutz, H., Furch, B.: Optical satellite communications in Europe. In: SPIE, vol. 7587, 758705 (2010)
13. Cheng, B.N., Charland, R., Christensen, P., Veytser, L., Wheeler, J.: Evaluation of a multihop airborne IP backbone with heterogeneous radio technologies. IEEE Trans. Mobile Comput. **13**(2), 299–310 (2014)
14. Zaouche, L., Natalizio, E., Bouabdallah, A.: ETTAF: Efficient target tracking and filming with a flying ad hoc network. In: Proceedings of the 1st International Workshop on Experiences with the Design and Implementation of Smart Objects, pp. 49–54. ACM Press, New York (2015)
15. Yanmaz, E., Kuschnig, R., Bettstetter, C.: Achieving air-ground communications in 802.11 networks with three- dimensional aerial mobility. In: Proceedings of the IEEE INFOCOM, pp. 120–124. IEEE, Turin (2013)
16. Abdulla, A.E., Fadlullah, Z.M., Nishiyama, H., Kato, N., Ono, F., Miura, R.: An optimal data collection technique for improved utility in UAS-aided networks. In: Proceedings of the IEEE INFOCOM, pp. 736–744. IEEE, Toronto (2014)

Design and Application Analysis of the Global Coverage Satellite System for Space Aeronautics ATM Information Collection

Changchun Chen[✉], Zhengquan Liu, Wei Fan, and Tao Ni

Shanghai Institute of Satellite Engineer, Shanghai, China
sc9@163.com

Abstract. This paper firstly analyzes the present situation and the demand of the current air traffic management surveillance, and then lead to the necessity of constructing the satellite system for space aeronautics air traffic management information collection. Combining with the needs of the current air traffic management surveillance, the design constraints about the satellite system for space aeronautics air traffic management information collection are analyzed. A Walker satellite constellation system constituted by 81 micro/nano satellites is therefore designed. Using STK numerical simulation tool, we can get the Earth which is covered by 100% and the maximum coverage gap time is not more than 40s. Finally, considering the capability of the satellite system for space aeronautics air traffic management information collection, the future application prospects of the system are discussed.

Keywords: Air traffic management surveillance · ADS-B · Micro/nano satellites · Walker constellation · Constellation efficiency · Inter-satellite links

1 Introduction

With the rapid development of commercial and general aviation, the timeliness and completeness of air traffic management surveillance data are becoming more and more demanding. Traditionally, air traffic management technology based on ground equipment such as the primary and the secondary radars is limited by the working mode and geographical location. It can-not meet the demand of air traffic management surveillance business for explosive growth, and is not suitable for the globalization of air traffic monitoring services. The urgent need for a new technical means to deal with the dilemma of current air traffic information management capacity.

The Automatic Dependent Surveillance-Broadcast (ADS-B) technology [1] realizes the real-time position information of the aircraft by the satellite navigation and positioning receivers installed on the aircraft, and is active to the ground and the surroundings by the airborne equipment. The aircraft broadcasts its own identification number, three-dimensional position and speed of flight, flight status and intent, and meteorological information, a technological means for the outside world to monitor it. Compared with radar and other technological means, ADS-B technology with higher data accuracy,

© Springer Nature Singapore Pte Ltd. 2017
Q. Yu (Ed.): SINC 2016, CCIS 688, pp. 266–273, 2017.
DOI: 10.1007/978-981-10-4403-8_23

faster data update rate of ICAO to develop the future of aviation surveillance technology, ADS-B is expected for the actual application all over the world in 2020. The working principle of the ADS-B technology is shown in Fig. 1.

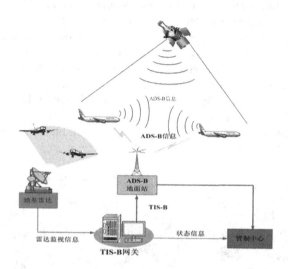

Fig. 1. The working principle of the ADS-B technology

At present, air traffic management technology based on ADS-B vehicle tracking technology mainly relies on land-based radar and still cannot overcome the limitation of land radar. The satellite system has a wide range of applications in communication, navigation, remote sensing and other fields with all-weather, all-weather, high spatial resolution and high temporal resolution. The ground plane ADS-B signal processing system is integrated into the satellite, you can achieve global aircraft operating status of real-time monitoring by the way through the multi-star network. At present, the United States, Europe, China and other major aerospace power carried by ADS-B satellites in orbit technology trials, is expected to have the ability to provide commercial services [2, 3]. From 2017 to 2020.

Considering the distribution of airplanes in the world, a quasi-real-time global aerial information acquisition system for air traffic management is designed in combination with the future application of aviation surveillance. The performance of the satellite system is analyzed by STK numerical simulation. Finally, the application prospect of the air traffic management information acquisition satellite system is analyzed according to the application demand and the future development trend.

2 Satellite System Design

2.1 The Analysis of the Design Constraints

According to the International Air Transport Association and ICAO statistics, land-based air traffic management system covers only 10% of the total airspace, and the remaining 90% or more airspace, especially as the ocean, desert and remote mountain airspace, where is no radar and ground ADS-B coverage. This seriously hampered the global air transport operations management efficiency and security capabilities. In addition, with the development of commercial aviation and general aviation in the future, the capability of ground air traffic management system is approaching saturation. It is urgent to establish an auxiliary means such as space-based air traffic management system to realize the integration of air traffic management system. Figure 2 shows the latitude distribution of global aircraft flight routes.

Fig. 2. The latitude map of the global aircraft flight route

For the above-mentioned needs of the global air traffic management, the design requirement of the regional coverage of the space-based airborne information collection satellite system is limited to the global airspace (90° between north and south latitude); because the data refresh rate of civil plane ADS-B signal is 1 s and the flight speed of the civil plane is about 900 km/h, the satellite system for space aeronautics air traffic management information collection area coverage interval design requirements are limited to less than 1 min.

Based on the above analysis, the coverage area of space-based airborne information collection satellite system is set to 90° north-south latitude and the time coverage interval is less than 1 min.

2.2 Satellite System Design

Space-based air traffic management information collection satellite system consists of multiple 50 kg-class micro-nano satellite. The ADS-B signal is received and processed by the payload in satellite. The ADS-B signal can be processed and forwarded to the ground station in real-time. To ensure real-time processing air traffic management information can be in a timely manner to the ground station, the satellite is equipped with

Ka-band inter-satellite communication payload, winch can be communicated within 5000 km distance in bi-directional 2 Mbps data transfer rate.

According to ADS-B signal bandwidth, the receiver noise figure, demodulation threshold and other parameters, to receive the aircraft ADS-B signal, the receiver sensitivity should be a minimum of −99.5 dBm. From the simulation results, we can see that the simulation results of the parameters such as the field radius, the link margin and the coverage area of different orbital heights. As shown in Fig. 3, when the satellite orbit altitude is 500 km, there are the maximum ground coverage and highest coverage efficiency for the ADS-B signal receiving payload (The coverage area ratio is the highest). Therefore, the orbit altitude of the air traffic management information collection satellite is selected as 500 km.

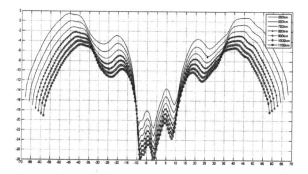

Fig. 3. Correspondence between the link margin of different orbits height and the radius of field of view

Corresponding to the orbital altitude of 500 km, in order to achieve a single satellite to the maximum coverage, the satellite antenna beam angle should be up to 68°. Using seven-beam array antenna to get the coverage area mosaic, Satellites can be approximated to ± 68° to cover the beam angle.

In order to achieve the global coverage of the design objectives, combined with the coverage of the beam angle of the payload, the orbital type of the air traffic management information collection satellite is the sun synchronous circular orbit, orbital inclination of 97.375° [4].

Considering the demand of inter-satellite data transmission in satellite system, the communication distance between the satellite and the inter-orbit satellite is not more than 5000 km. Therefore, The number of satellites in the same orbital plane is not less than 9 and the number of orbital planes is not less than 9 [5, 6].

In summary, the constellation plan of the satellite system for space aeronautics air traffic management information collection is the Walker constellation composed of 81 satellites, and the 81 satellites are distributed on 9 orbital planes. Constellation diagram of the configuration and coverage of the ground diagram is shown in Figs. 4 and 5.

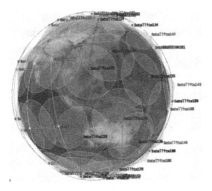

Fig. 4. Schematic diagram of satellite constellation configuration

Fig. 5. Schematic diagram of satellite system ground coverage

3   System Effectiveness Analysis

3.1   System Effectiveness Analysis

With the STK simulation software, the global coverage of the space-based air traffic management information collection satellite system is simulated, the simulation results are shown in Fig. 6. In Fig. 6, the x-axis represents the latitude that the satellite system can cover, and the y-axis represents the maximum, minimum, and average coverage

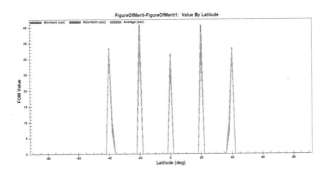

Fig. 6. Satellite system performance coverage analysis chart

intervals over different latitudes through the satellite system. From the simulation results, we can see that the satellite system can achieve the continuous coverage of the global interval with the maximum coverage interval of 40 s, which satisfies the system design constraints.

3.2 System Inter-Satellite Link Analysis

Since all satellites in the Walker constellation are geometrically symmetric, the characteristics of any satellite may represent the geometrical characteristics of all the satellites in the constellation. Therefore, the inter-satellite link analysis in satellite constellation, you can choose any one of the constellation satellite inter-satellite links to analyze the situation. In this paper, the first orbital plane of the constellation system is selected and the first satellite, Satellite 11, is used as the representative satellite to analyze the inter-satellite communication link between the satellite 11 and the neighboring satellites Satellite 12, Satellite 19 and the adjacent satellites Satellite 21 and Satellite 99.

Using STK simulation software, the communication range, Azimuth and Elevation of satellite data communication link between Satellite 11 and adjacent Satellite 12 and Satellite 19 in the same orbital plane are analyzed respectively, as shown in Figs. 7 and 8. The simulation results show that the distance between the Satellite 11 and the neighboring Satellite 12 and Satellite 19 in the same orbital plane is less than 5000 km, which can realize the real-time data transmission between the adjacent satellites in the same orbit.

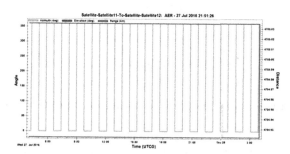

Fig. 7. The analysis chart of Inter-satellite links between Satellite 11 and Satellite 12

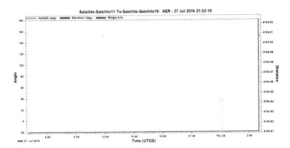

Fig. 8. The analysis chart of Inter-satellite links between Satellite 11 and Satellite 19

The communication range, Azimuth and elevation of satellite data communication link between Satellite 11 and adjacent Satellite 21 and Satellite 99 are analyzed by STK simulation software, as shown in Figs. 9 and 10. Simulation results demonstrate that Satellite 11 and adjacent Satellite 21 and Satellite 99 have a distance of 5000 km between them, which can realize the real-time satellite data transmission between adjacent orbital planes. Therefore, the inter-satellite links of the satellite system are interoperable which can meet the design requirements that the data can be back to the ground station in time.

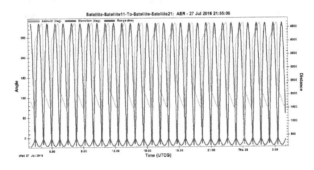

Fig. 9. The analysis chart of Inter-satellite links between Satellite 11 and Satellite 21

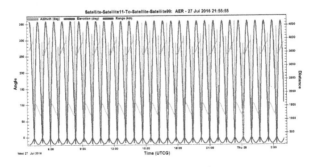

Fig. 10. The analysis chart of Inter-satellite links between Satellite 11 and Satellite 99

4 System Application Domain Analysis

Global coverage of air traffic management information acquisition satellite system can achieve worldwide plane ADS-B signal real-time reception and processing, and its application areas are mainly in the following areas:

(1) global aircraft safe operation state monitoring and track optimization;
(2) aircraft channel operational situation information awareness and broadcasting;
(3) the global airport flight landing auxiliary scheduling management;
(4) global air cargo management and economic situation analysis;
(5) air crash rescue, rescue information data support.

5 Summary

The demand of air traffic management information monitoring service is becoming more and more urgent, which makes it necessary to build the satellite system for space aeronautics air traffic management information collection for global real-time coverage. In this paper, a space-based airborne information acquisition satellite system consisting of 81 micro-satellites is designed. The design constraints and constellation design of the system are analyzed, and the rationality of the satellite system design is verified by numerical simulation. Combining with the ability of air traffic management information collecting satellite system, the potential application field of the system is analyzed.

References

1. Li-Hu, C., Xiao-Qian, C., Yong, Z.: The onboard ADS-B receiver system and its applications. Satellite Application **2016**(3), 34–40 (2016)
2. Yuan, X.-J., Zhang, J., Huang, Z.-G.: ADS-B clustering algorithm for air-based and satellite–based combined surveillance system. Telecommun. Eng. **41**(1), 82–85 (2007)
3. Zhang, Q.-Z., Zhang, J., Liu, W., Zhu, Y.-B.: Investigation for main problem of ADS-B implementation in ATM. Meas. Control Technol. Instrum. **9**, 72–74 (2007)
4. Zhang, Y.-L., Fan, L., Zhang, Y., Xiang, J.-H.: Theory and Design of Satellite Constellations. Science Press, Beijing (2008)
5. Xia, Y., Li, J.-C.: Inter-satellite links analysis of walker constellation. J. Geodesy Geodyn. **32**(2), 143–147 (2012)
6. Shuang, X., Wang, X.-W., Min, H., Ma, L.-B.: An intelligent analysis method of satellite network capacity. Chin. J. Comput. **2016**(39), 1–11 (2016)

A Distributed Algorithm for Self-adaptive Routing in LEO Satellite Network

Hao Cheng[1], Meilin Liu[2], Songjie Wei[1(✉)], and Bilei Zhou[2]

[1] School of Computer Science and Engineering, Nanjing University of Science and Technology, 200 Xiaolingwei, Nanjing 210094, China
swei@njust.edu.cn
[2] Shanghai Institute of Satellite Engineering, Shanghai Academy of Spaceflight Technology, 3666 Yuanjiang Rd, Shanghai 201109, China

Abstract. LEO satellite networks, represented by the successful Iridium System, are composed of multiple satellite nodes and inter-satellite links (ISL). Numerous routing algorithms have been designed to determine satisfying routes between data flow sources and destinations, within the constraints including delay, congestion control, and throughput and load balancing. This paper proposes a distributed network-state aware self-adaptive routing algorithm based on neighbor ISL status and node workload. Every satellite node is independently responsible for forwarding datagrams in its queue, with information about network status piggybacked in the transmitted datagrams. Such information helps understand and predict the network workload status on each direction of the outgoing links, and is used for nearly-optimal selection of datagram outbound links to achieve load balancing and multi-path routing. Experiments are conducted on ns-2 simulation platform with a designed LEO walker, to implement and evaluate the effectiveness and efficiency of the proposed algorithm. The results show a significant improvement of more than 50% on the network workload balancing, with a few more hops in the selected multiple routing paths compared with the traditional Dijkstras shortest path algorithm.

Keywords: LEO · Satellite network · Routing algorithm · Self-adaptive

1 Introduction

In 1965, Instelsat-1 was launched as the first commercial communication satellite to provide instantaneous cross-continent telecommunication service. After that, communication satellites are extensively applied in all aspects of human civilization advance and daily life, in military, economy, culture and society development in the past 50 years. In terms of quantity, communication satellites make up of the largest category of artificial satellites in space. Constrained by orbit, payload capability, communication coverage, running time and system stability, a single satellite cannot cover the whole earth to supply realtime service continuously. With the development of space technology and application, satellites in

© Springer Nature Singapore Pte Ltd. 2017
Q. Yu (Ed.): SINC 2016, CCIS 688, pp. 274–286, 2017.
DOI: 10.1007/978-981-10-4403-8_24

different orbits are deployed and organized as constellation. Satellites in a constellation coordinate for larger ground coverage, operate together under shared control, synchronize for more complicated functionality. Amid the modern computer technology, network protocols and control theory, multiple satellites in different orbits can be coordinated for a full coverage of the Earth for a seamless intelligent information service, i.e., 'a hybrid of space and terrestrial information network' for data relaying and information sharing. Every satellite node in the network is independent and capable of sending, receiving and forwarding data. Inter-Satellite Link (ISL) for data communication is maintained between neighbor nodes in constellation. Datagram is transmitted between satellites in specific paths called routes [1]. It is of great value to keep improving and optimizing satellite routing algorithms for promoting the development of space communication and optimizing space information resource usage. Efficient inter-satellite routing strategy improves the data transmission rate, QoS and system reliability in satellite internetworking.

Packet routing is the fundamental functionality on the network layer in TCP/IP protocol suite. A routing algorithm determines how a data packet is transmitted along switches in network. Modern Internet based on TCP/IP employs hierarchical routing by choosing different interior and exterior gateway protocols to implement data switching in network. On the other hand, the topology dynamics and the asymmetric links in satellite network devalue a direct application of the traditional TCP/IP protocols in satellite communication scenario [2]. Other researchers refer and extend the sophisticated MANET routing protocols such as OLSR and AODV, and have proposed several multi-path, dynamic and load-balanced routing algorithms for satellite constellation. Satellite network is fundamentally different from MANET. Mesh nodes usually have isotropic communication capability. Data links in satellite network are influenced by the relative movement and angles between peers. Fast-changing relative distances among satellites result in nonnegligible communication delay fluctuation and link instability, which is a big challenge for those mesh routing protocols to run effectively in satellite constellation [3].

The satellite technology research communication and industry have not come up with a widely acceptably and universally applicable routing protocol for satellite internetwork. Existing candidates, either from the mature TCP/IP, the fast-pacing MANET or the specific CCSDS [4] series, do not fully take account of the asymmetric satellite network traffic scenarios and the uneven data flow distribution. They have rigorous prerequisites for computation power, storage and transmission capabilities on satellite nodes. Many tend to oversimplify the requirements on satellites for stability and robustness.

This paper proposes an self-adaptive datagram relay and routing decision algorithm for LEO satellite constellation. LEO satellites have significant advantages on less time consumption and lower economic cost in design, manufacturing, launching and networking procedures. LEO is suitable for various communication and surveillance missions. It is a trend for LEO network to conduct routing based on inter-satellite link (ISL) independent of ground-based assistance.

A practical LEO inter-satellite routing strategy must implement automatic decision-making, self-adaptive, network autonomous without relying on stationary network topology, satellite orbit, pre-configured routing path, additional assistance from ground.

Although all satellite node in a constellation can be organized into a network to collaborate on missions, ideally every node should independently compute routing decisions and conduct forwarding operations based on only its own perception of network connection and traffic status. Inspired by the cooperative game theory with incomplete information in economics, referring to the Internet Autonomous System (AS) model, we propose an self-adaptive routing framework in which satellite node are autonomous, distributed and independent in sensing network status and making datagram forwarding decisions. We use the network flow strength state as an example to demonstrate how the proposed approach may help improving network-wide workload balancing and flow distribution. The advantages of the proposed routing algorithm can be summarized as follows:

- Distributed routing system. Routing decision is dispersedly made on every node in satellite network. There is no centralized routing coordinator or global routing information. Node routing function is distributional and parallel, avoiding active routing discovery, cooperation and maintenance between nodes.
- Autonomous satellite node. Every satellite node is responsible for datagram forwarding queued locally, calculating routing path independently, and sensing network and link status continuously.
- Self-adaptive routing strategy. Self-adaptive routing is implemented for global convergence through sensing network status and choosing various network performance indicator according to different application scenarios and service requirements.
- Scalable network structure. Proposed network model is scalable when a single node is added, disconnected, overwhelmed or abused.

2   Self-adaptive Inter-satellite Routing

Satellite constellation is a set of satellites according to specific organizational rules. Design of constellation determines the topology of satellite network. There are two kinds of LEO satellite network structures: polarized and inclined orbit constellations. They differentiate in orbit inclination angle and radian distance distribution in equatorial plane. A satellite network system is naturally divided into two hierarchies: satellite nodes and planes, similarly to hosts and ASes in Internet [5].

2.1   Satellite Node and Network Model

We first describe the satellite network model which includes node model, link model, and network flow model. An ISL between neighbor satellites in the same plane(orbit) is an intra-plane link, while one connecting cross-plane(cross-orbit) satellites is an inter-plane link. Every satellite maintains inter-satellite links with neighbors in local and adjacent orbits. A single satellite is modeled to have 4 ISLs, illustrated in Fig. 1. For example, satellite $S22$ has links with $S21$, $S23$ as *up_link* and *down_link* inside the same orbit plane. $S22$ also has $S12$ and $S32$ to its left and right neighbors in adjacent orbits, respectively as *left_link* and *right_link*. The relative inter-satellite distance and angle is constant inside the same orbit, so that intra-plane ISLs can be built and maintained continuously. The relative distance, speed and angle of satellites are time varying between difference orbits. Cross-plane ISLs need special procedure on link state detection and maintenance. In particular, for polarized satellite constellation, there is a seam when two satellite planes run in opposite directions, such as in Fig. 1 between *orbit1* and *orbit4* when the topology spreads on sphere surface [6]. The cost of building and maintaining trans-seam links is very expensive, and will complicate the 2D structure in Fig. 1 into a 3D structure. Previous research indicates that trans-seam links only have trivial impact on network performance when doing satellite networking and communication [7].

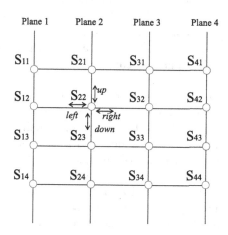

Fig. 1. Network topology with nodes and links in satellite constellation

Every satellite node keeps tracking the traffic workload state of its each ISL. Link state as traffic statistics includes inbound/outbound traffic, transmission delay, connection time and stability, etc. Different Network QoS optimization goals can be achieved when balancing routing strategy and network traffic by choosing the corresponding nodes and link performance. The link traffic workload is selected as a performance measure in the following discussion of this paper as an example, and to optimize routing for data traffic sensing and load-balancing between satellite nodes. Every satellite node S_i records inbound and outbound traffic on link j as $L_{i,j}^{in}$ and $L_{i,j}^{out}$.

Communication satellite nodes are responsible for data transmission service by routing and forwarding datagrams between data flow sources(senders) and destinations(receivers). Flow can be initiated and consumed by various end nodes such as spacecrafts, gateways, ground fixed and mobile devices. Every data traffic flow F_k enters and exits the network via border satellite nodes in the network. This paper only focuses on satellite network routing, and thus separates networking nodes from traffic end nodes. Without losing the generality and faithfulness, we simplify data traffic as each flow k initiated from the entrance node F_k^{src}, and sunk at the exit node F_k^{dst}.

2.2 Design Assumptions

Before presenting the proposed link-state sensing self-adaptive routing algorithm on independent satellite nodes, we first discuss and argue a few assumptions about the satellite nodes, inter-satellite links, network structure, and the traffic datagram. While keeping in a realistic and practical scope of satellite networking construction and maintenance, these assumptions simplify the algorithm design complexity and thus makes the presentation clearer with the points.

We first assume that every satellite node has some capability of link-state sensing, data computation and storage, and constellation awareness. Link-state sensing means a satellite is aware of its links to neighbor nodes both intra and inter orbit planes. A satellite also has enough computation and storage resource for routing. As a member of a constellation, each satellite has a global view of the constellation orbit planes and satellite members, including numbers, orbits and movement parameters.

Although the link data transmission rate can be asymmetrical, every ISL can transmit data bidirectional. ISL connection lasts longer than the data transmission delay on it. A satellite node or ISL may fail. A malfunctioning ISL cannot transmit data between the satellite nodes, and a malfunctioning satellite node losses all of ISL connections it has with neighbors.

A data flow in satellite network is transmitted as a sequence of datagrams. Every datagram has a fixed size and a predefined structure. Its content organization is mutually agreed between the data sender and receiver according to a communication protocol. Forwarding nodes in network can repackage datagram by adding envelop head and updating datagram piggback info. Datagram specifies its targeted receiver node and the receiver's orbit. The transmission process of datagram is similar with Internet UDP datagram transmission which is not connection-oriented, with no guarantee on datagram delivery, integrity, timeliness, and order.

2.3 Sensing and Exchanging Network Congestion

Every satellite node needs sense its ISL status, add extra information in passing-by datagram to share node workload status, infer load-balance of neighbors in network based on aforementioned information. All of those information are integrated as an guidance when choosing optimal next hop ISL to forward datagram. As described in last session, datagram piggybacks node traffic statistics. By extending the piggybacked information to cover other measures like queuing delay, link latency, buffer availability, etc., it is possible to apply the same routing strategy on satellite network to achieve different QoS optimization objectives.

A satellite node S_i maintain possible ISLs to neighboring nodes. For each link, the node tracks two statistic records namely $L_{i,j}^{in}$ and $L_{i,j}^{out}$ which count inbound and outbound traffic, respectively. Here $j \in \{left, right, up, down\}$. The maintained datagram load on the corresponding link is linearly increased as below when time passes:

$$L_{i,j}^{\{in,out\}} = L_{i,j}^{\{in,out\}} \cdot aging\_ratio + c \cdot N \tag{1}$$

where N is the number of datagram; c is a non-negative growing rate predefined depending on the type, size and upper-layer protocol of traffic datagram. Since L accumulates gradually, we apply an aging factor onto L when time elapses. $aging\_ratio$ values between 0.0 and 1.0 as a float. It is positively correlated with the size of network and transmission delay between nodes.

A datagram P piggybacks information about the workload of all passed-by nodes on its route. This information is inserted by the sender node, updated and consumed by the forwarding nodes, and removed by the receiver node. We use $Load_p^{plane}$ to denote the statistics and the estimate of the datagram load on nodes of current orbit plane. If satellite S_i is the current node processing this datagram P, and it chooses ISL j as an outbound link, then $Load_p^{plane}$ is calculated as

$$\begin{cases} Load_{S_i} & \text{if } S_i \text{ is a sender} \\ Load_{S_i} + hop_{aging} \cdot Load_p^{plane} & \text{if } S_i \text{ is forwarding to intra-plane ISL} \\ Load_{S_i} & \text{if } S_i \text{ is a fowarding to inter-plane ISL} \end{cases} \tag{2}$$

hop_{aging} is the decay rate as hop count increases when datagram travels along orbit. If there are M nodes on the same orbit plane without loop, a properly selected decay rate should achieve $(hop_{aging})^{\frac{M}{2}} \Rightarrow 0$.

If the other side of the selected ISL j is the satellite node S_k, then once the datagram arrives at S_k, the inbound traffic statistic should be updated as

$$L_{k,j}^{in} = L_{k,j}^{in} \cdot aging\_ratio + c \cdot Load_p^{plane} \tag{3}$$

Here we interpret $Load_p^{plane}$ as the accumulated traffic load on all the passing-by intra-plane nodes right before the datagram is leaving node S_i to the next same-plane node. This is also how much traffic the next-hop node may expect maximally from the upstream along this link. When the next hop is on a different orbit plane, $Load_p^{plane}$ represents an estimate of the traffic load on the current orbit plane.

2.4   Packet Forwarding on Satellite Node

When a datagram arrives at satellite S_i, this node as a network switch starts to compute independently for a forwarding operation. Figure 2 explains the procedure and the decision strategy. If the current node is the dedicated datagram receiver, then it delivers the datagram to the above-layer application and finishes. If the destination is on the same orbit plane as S_i, it picks an intra-plane ISL as an outbound link. For datagram heading to a destination on a different plane, the next hop could be either of the four neighbor nodes connected with S_i. If there is more than one candidate appropriate as outbound link for the datagram, S_i chooses by comparing the sum of $L^{in} + L^{out}$ value as the link workload. ISL with the minimum cumulative workload wins to carry on the datagram.

2.5   Avoiding Loops in Routing

With the layout and organization of satellite nodes in a constellation, ISLs within and between orbit planes may form various routing loops. Any effective routing algorithm should avoid an intermediate node to receive any datagram more than once, i.e., a loopback in routing. Figure 3 shows all the possible inter-satellite links in a 120/10/1 LEO constellation. Close loops are formed in every orbit plane and across all the planes. Since we are working on forwarding datagram in a distributed routing system without centralized coordinator or monitor node, no satellite itself can either detect or avoid loop in routing paths alone. Special regulation and treatment is obligatory to avoid routing loops when conducting the proposed self-adaptive distributed routing algorithm in LEO satellite network.

Setting and checking Time-To-Live (TTL) value is a light-weighted but effective mechanism in avoiding infinite forwarding loop. In each datagram as supplement information, send node initializes two positive TTL values namely $TTL_{overall}$ and TTL_{plane}. The former is the maximum number of satellite hops the datagram may traverse before it either reaching the destination or getting dropped. The initial value is determined based on the network size, the node resource richness, and the application service type of the satellite constellation. The latter defines the most number of intra-plane forwarding that a datagram may experience before it proceed to the next orbit plane. Every forwarding node should deduct these two TTLs by 1 before sending the datagram to an outbound link. TTL_{plane} is restored to the initial value when a datagram is forwarded across orbit plane. During the forwarding procedure in Fig. 2, if both inter- and intra-plane links are available, the forwarding satellite will choose intra-plane link with a priority proportional to the remaining value of TTL_{plane}. With a zero value of TTL_{plane}, unless there is no inter-plane satellite link available, S_i is forced to forward this datagram to a neighbor satellite on an adjacent orbit.

As a special case of loops in routing, another unexpected scenario is when a datagram is forwarded back-and-forth between two adjacent nodes. In Fig. 1, it is possible that after computing the forwarding link priority based on the procedure in Fig. 2, satellite S_{22} chooses the downlink to S_{23} as local routing optimal, which happens to repeat the procedure and determines the uplink to S_{22} is the best choice locally. Such locally back-and-forth shifting will of course be terminated once the TTL value is exhausted, or when the workload on S_{22} or S_{23} changes after some iterations, but still it brings in extra routing delay, unnecessary resource consumption, and meaningless TTL deduction. To deal with this case, we insert one more complement parameter $D_p^{plane} \in \{0, 1, -1\}$ in

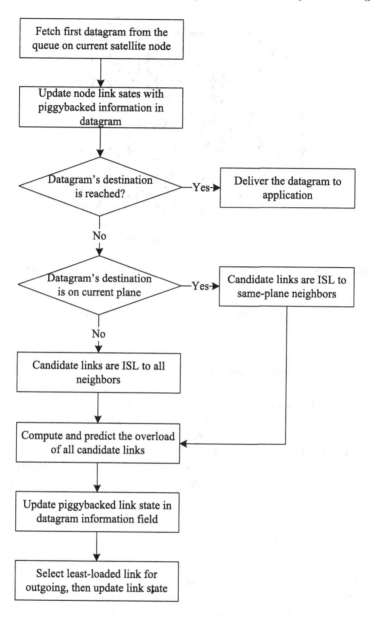

Fig. 2. Datagram forwarding procedure on satellite node

a datagram. This parameter flags the direction in constellation that this datagram is traveling. A value of 1(or −1) indicates the datagram is traveling in the up(or down) direction along the current orbit. On the first hop in current plane, $D_p^{plane} = 0$ means the intra-plane direction is not determined yet. The satellite may freely forward the datagram to either direction and set the $D_p^{plane} = 0$ value.

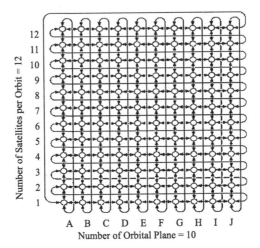

Fig. 3. Logical connections in a constellation with inter-satellite links [6]

This value is reset to 0 when cross-orbit forwarding happens. Also only the next orbit plane closer to the targeted orbit is selected for inter-plane forwarding.

3   Experiments and Results

To validate the effectiveness and reliability of the proposed routing strategy, we first use the Satellite Toolkit software (STK) to design a polarized orbit constellation similar the Iridium system. STK exports the orbit, connection status and continuous time parameters into ns-2 simulator. We extend the routing decision module in ns-2 to implement the proposed link-state sensing, information piggyback in datagram, and routing strategy. Two different application scenarios are configured for simulation experiments.

3.1   Simulation Configuration

A walker star constellation is designed using STK which consists of 6 circular orbits of height 680 km and inclination angle 84°, 54 satellites, to covers the whole Earth surface. A 3D demonstration of the constructed constellation is shown in Fig. 4. The TTL_{plane} and $TTL_{overall}$ are set as 8 and 32, respectively. $aging\_ratio$ and hop_{aging} are 0.98 and 0.5 each.

3.2   Network Traffic Scenario

To evaluate the effectiveness and stability of proposed adaptive routing algorithm, multiple network traffic flows are configured with source and destination nodes distributed across the whole network. The simulation runs for 100 s with 1000 randomly-generated network traffic flows. The traffic sender/receiver nodes

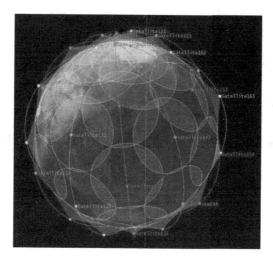

Fig. 4. 3D overview of the simulated satellite walker

are choosing randomly in 54 satellite nodes to make the total traffic flows follow lognormal distribution [8]. This means a few nodes in network initiate and sink most datagrams, and most others nodes only involve by relaying datagrams. Flow sending rate is defined as the number of datagrams sent from traffic source node per second, and it follows normal distribution with a mean of 100. Every data flow lasts 10–20 s.

3.3 Results and Discussion

Two performance indicators are measured in the experiments. First, the length of inter-satellite routing path is calculated as the hop count in satellite network. Routing hops indicate that the routing overhead of a datagram during its path from source to destination. Fewer hops implies less transmission delay and thus possibly less time delay between the two ends which is composed of queue and routing delay, propagation and transmission delay. The proposed adaptive routing algorithm in this paper utilizes link-state awareness to dynamically distribute traffic flows. Datagrams from the same flow may take multiple paths besides the shortest path. So the hop count increases as the extra routing cost in system, i.e., extra number of nodes passed by. Second, the distribution of traffic datagrams in network is measured as the number of datagrams queued, processed and forwarded on different node at the same time. It is quantified as the variance of this number. Only nodes with inbound traffic in the past second are counted.

In order to benchmark the performance of the proposed routing algorithm, we also test the above two measures with the shortest-path routing simulated. Figures 5 and 6 presents the simulation results and comparisons. Flows using the short-path routing traverse 7 relay nodes on average. While with the proposed self-adaptive routing, Fig. 5 shows similar hop counts at the beginning,

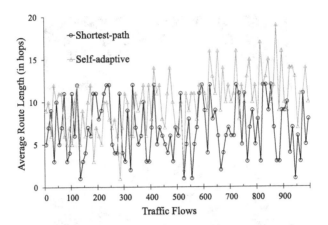

Fig. 5. Length distribution of inter-satellite routing paths

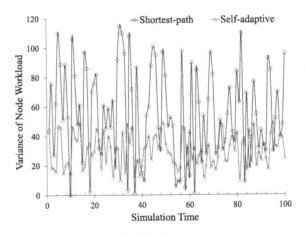

Fig. 6. Workload distribution of inter-satellite flow traffic

but dynamically self-adjust during the simulation when more flows and traffic join in. With more information about link-state is collected and exchanged gradually, satellite nodes in the proposed routing tend to choose outbound link to the less overwhelmed neighbors and planes. This increases the average flow hop count by 2–3 to 7–10 hops. In a realistic LEO satellite network, ISL transmission delay is in the magnitude of tens of milliseconds, but the node queuing and processing delay may be as large as hundreds of milliseconds or more, due to less powerful computation and onboard storage capability. So the increased number of hops do not necessarily lead to a growing end-to-end (source-to-destination) flow transmission delay, as long as idle and faster satellite hops are preferred.

Figure 6 shows how the traffic workload is distributed and balanced among multiple satellites in network. The shortest-path based routing, when handling non-uniformly distributed traffic flows which are typical in practice, is more likely to cause congestion in some pivotal satellites nodes and links, especially those inter-plane links [9]. In this case, workloads on satellites are dramatically differentiate without being balanced, with the variance measure fluctuating between 0 and 120. Using the proposed self-adaptive routing, workload in number of datagrams processed on each satellite is balanced with the variance more stable in 0–50, nearly 50% improvement in load balancing.

Measures of load balancing from other prospects such as network flow delay, congestion spread and traffic loss ratio are also simulated, with similar results observed supporting the superiority of the proposed algorithm over shortest-path routing. We skip the result illustrations due to page limit.

4   Conclusion

LEO communication satellite systems, such as Iridium, Globalstar and Teledesic, provide global connection and seamless coverage. They represent the future direction of satellite communication technology and application. A single LEO satellite has limited communication service capability because of its insufficient orbit height. Multiple LEO satellites can form constellation to break such limit. Networking and routing is one of the critical problems to be solved before LEO satellites can coordinate to provide reliable, economical and efficient communication service.

We propose a distributed self-adaptive routing strategy for LEO satellite in network. Each satellite node by tracking and estimating the traffic workload in neighbor nodes and orbit planes, independently prioritize nodes and planes with more resources and capabilities. By transmitting flows on multiple routes, the proposed algorithm can gradually self-adjust routing on each satellite to achieve network-wide node workload balancing. Experiments with ns-2 simulation demonstrate the effectiveness of the new routing algorithm, and provide a framework to be extended for optimizing network transmission performance measured in various aspects and expectations.

Acknowledgments. This material is based upon work supported by the CASC Innovation Fund No. F2016020013, and the CERNET Next-Generation Internet Innovation Project under contract No. NGII20160601. Opinions and conclusions expressed in this material are those of the authors and do not necessarily reflect the views of the sponsors.

References

1. Lu, Y., Zhao, Y., Sun, F., et al.: Routing techniques on satellite networks. J. Softw. **5**, 1085–1100 (2014)
2. Papapetrou, E., Pavlidou, F.: Distributed load-aware routing in LEO satellite networks. In: IEEE Globecom. IEEE Press (2008)
3. Li, J., Li, D., Wang, G.: New dynamic routing algorithm based on MANET technology in the LEO/MEO satellite network. J. Commun. **26**(5), 50–56 (2005)

4. Consultative Committee for Space Data Systems (CCSDS): Space Internetworking Services Area Publications https://public.ccsds.org/Publications/SIS.aspx
5. Liu, G., Li, H.: Satellite Communication Network Technology. Posts & Telecommunications Press, Beijing (2015)
6. Suzuki, R., Yasuda, Y.: Study on ISL network structure in LEO satellite communication systems. Acta Astronaut. **61**(7–8), 648–658 (2007)
7. Gavish, B., Kalvenes, J.: Impact of intersatellite communication links on LEOS performance. Telecommun. Syst. **8**(2), 159–190 (1997)
8. Camps, F., Thibault, E., Harasse, S.: Statistical distribution of traffic sources in network simulation tools. In: World Congress on Engineering and Computer Science (2008)
9. Li, C., Liu, C., Jiang, Z., et al.: A novel routing strategy based on fuzzy theory for NGEO satellite networks. In: Vehicular Technology Conference (2015)

Author Index

Cao, Lifeng 137
Chang, Chengwu 80
Chen, Changchun 266
Chen, Dong 172
Chen, Qianbin 126
Chen, Wanli 117
Chen, Xiang 117
Cheng, Hao 274
Cheng, Liyu 80
Cheng, Zijing 16
Cui, Gaofeng 151
Cui, Zhaojing 16

Dai, Cuiqin 126
Di, Xiaoqiang 182
Du, Xuehui 137

Fan, Jing 80
Fan, Wei 266

Gao, Jian 232
Gao, Jingchun 95
Gao, Zihe 172
Guo, Lei 126

He, Chen 202
He, Yizhou 151
Huang, Huiming 35, 232
Huang, Xiangyue 244
Huang, Xiujun 54

Jia, Yun 3
Jiang, Huilin 182
Jiang, Lingge 202
Jiang, Yong 66, 256
Jiang, Yuming 182

Kuang, Linling 35

Li, Jinqing 182
Li, Wei 16
Li, Yongjun 66, 256
Lian, Zhuxian 202

Liao, Ying 137
Liu, Chunmei 162
Liu, Hongyang 80
Liu, Jincan 45
Liu, Kai 117
Liu, Kaiming 95
Liu, Meilin 274
Liu, Xiang 16
Liu, Zhengquan 266
Lu, Xingpei 172
Luo, Bin 232

Ma, Haiquan 16
Ma, Zongfeng 26, 54
Meng, Xiangjun 244
Miao, Ye 16

Ni, Tao 266

Qu, Zhichao 26

Ren, Guangliang 104

Shen, Jingshi 3
Shen, Yufei 172
Shi, Dele 54
Si, Chaoming 217
Song, Qingyang 126
Sun, Lichao 26

Tao, Ying 172

Wang, Cheng 151
Wang, Houtian 172
Wang, Jingchao 45
Wang, Lei 35
Wang, Min 182
Wang, Qiwei 104
Wang, Weidong 151
Wang, Xiang 256
Wei, Songjie 274
Wen, Guoli 172
Wu, Jueying 104

Xia, Lei   217
Xiao, Shaoqiu   162
Xie, Gang   95
Xin, Mingrui   3

Yan, Jian   35
Yang, Huamin   182
Yi, Zhuo   137

Zhang, Honggang   244
Zhang, Keke   217
Zhang, Shengyu   217

Zhao, Chao   95
Zhao, Haiyan   66, 256
Zhao, Jing   256
Zhao, Shanghong   66, 256
Zhao, Zhifeng   244
Zheng, Yongxing   66
Zhou, Bilei   274
Zhou, Hongbin   45
Zhou, Shihong   66
Zhou, Shilong   217
Zhu, Liang   232

Printed in the United States
By Bookmasters